TEA CULTIVATION IN THE PLAINS OF NORTH EAST INDIA

THE AUTHORS

Dr. Sayed Edward Kabir (b. 3rd Feb.1954) presently serving as an Assistant Professor in the Department of Tea Science, University of North Bengal. He was involved in tea research at Darjeeling Tea Research Centre under Tea Board, India from 1978 to 1996. He is a Technical Consultant of International Institute of Biotechnology and Toxicology (IIBAT).

Dr. Abhaya Prasad Das (b. 8th May 1951) is a Professor in the Department of Botany and former Head of Tea Science Department of North Bengal University. Prof. Das has established East Himalayan Society for Spermatophyte Taxonomy.

TEA CULTIVATION IN THE PLAINS OF NORTH EAST INDIA

–Authors–

Dr. S. E. Kabir

Dr. A. P. Das

2021

Regency Publications

A Division of

Astral International (P) Ltd

New Delhi 110 002

ISBN: 9789351306955 (Hardbound)

Published by :

Regency Publications
A Division of
Astral International Pvt. Ltd.
– ISO 9001:2015 Certified Company –
4736/23, Ansari Road, Darya Ganj
New Delhi-110 002
Ph. 011-43549197, 23278134
E-mail: info@astralint.com
Website: www.astralint.com

राजीव प्रताप रुडी
RAJIV PRATAP RUDY

कौशल विकास एवं उद्यमिता राज्य मंत्री
(स्वतंत्र प्रभार) एवं
संसदीय कार्य राज्य मंत्री, भारत सरकार
MoS Skill Development & Entrepreneurship
(Independent Charge) &
Parliamentary Affairs, Government of India

18th December, 2014

Message

The book entitled " Tea Cultivation in the Plains of North East India" by Dr. Sayed Edward Kabir and Dr. A. P. Das of the Department of Tea Science, University of North Bengal is a suitable book for upliftment of Tea production and quality of Indian Tea. Many important aspects have been discussed in this book with plenty of Illustraitons. This book will certainly serve as a useful handbook for the Tea planters of India. I appreciate the effort of the authors in bringing out this book

Rudy

Rajiv Pratap Rudy

Office: Room No. 56, Parliament House, New Delhi-110001 Tel : 011-23010895, 23034638, Fax: 23011824

Residence: AB-97, Shahjahan Road, New Delhi-110011 Phone: 011-23070300, 23070999 (Telefax)

PREFACE

Tea is a multifaceted subject. Tea being a highly important commodity in India, there is need to improve the productivity as well as quality of tea with the introduction of scientific methods of cultivation. This book endeavours to present a practical aspect of garden practices with few advice. The protocols and methodologies exercised in this book is completely based on the practical experience of the authors and inputs from various Tea Research Centres like Tocklai Experimental Station, UPASI Tea Research Foundation etc.

The content of this book may not conform to the findings of the Tea Research Centres or other organizations. It is not at all obligatory on the part of the tea gardens / tea planters or any body else to follow the instructions and advices provided in this book.

Many pictures and descriptions have been taken from various books, which have been duly acknowledged. Publications of Tocklai Experimental Station , Jorhat and UPASI Tea Research Foundation, Valparai, which are the most renowned Tea Research Centres in the world are the main strength of this book. The authors cordially and humbly acknowledge their support in bringing out this book.

Products (*viz.* pesticides, growth promoters etc.) of many companies have been included in this book. The authors tried to accommodate almost all the products available in the market , however, there may be many more products which could not be included. The authors assure that if any omission if reported will be accommodated in the next edition.

The authors are indebted to Dr. N.Muraleedharan, Director, Tocklai Experimental Station, Dr. B.Bera, Director (Research) , Tea Board, Dr. A.Babu, Deputy Director (Research) & Senior Principal Scientist, In-Charge of NBRRDC, Dr. F.Rahman, Tea Scientist, Dr. S.K.Debnath, Tea Scientist , Md. Fareed, Senior Advisory Officer & Principal Scientist, Tea Research Association for their immense contribution to this book. The authors express deep sense of gratitude to Prof. Indrajit Ray, Vice-Chancellor of Thakur Panchanan Barma University, Coochbehar and Prof. Subhash Chandra Santra of Kalyani University for their inspiration in publishing this book. Authors are grateful to T. Stanes & Company, Coimbatore for providing inputs and photographs. Authors express deep sense of gratitude to Mr. Mainak Mukherjee, Department of Tea Science, North Bengal University for helping with notes on Soil Science. Authors are also grateful to S.B.Agency, Siliguri for providing many data on pesticides etc. The role of Mr. Siddhatha Hazra , Miss Priyanka Banerjee and few other students of M.Sc. in Tea Science Course in the North Bengal University in writing this book is immense.

This book is supposed to serve as a tool for the tea planters, especially the new comers in the field of tea and the students pursuing tea courses. The authors have tried to make the book as simple as possible and avoid too much of technicality. It can serve as a practical hand book for the tea planters.

S. E. Kabir

A. P. Das

CONTENS

1

HISTORY OF TEA AND TEA VARIETIES

Geographic Origin of Tea

Camellia sinensis originated in southeast Asia, around the intersection of latitude 29°N and longitude 98°E, the point of confluence of the lands of North East India, north Burma (now Myanmar), southwest China and former Tibet . The plant was introduced to more than 52 countries, from this 'centre of origin'," rather than as a medicinal concoction".

Discovery of Tea

In one popular Chinese legend, Shen Nung, the legendary Emperor of China and inventor of agriculture and Chinese medicine was drinking a bowl of just boiled water due to a decree that his subjects must boil water before drinking it sometime around 2737 BC when a few leaves were blown from a nearby tree into his water, changing the colour. The emperor took a sip of the brew and was pleasantly surprised by its flavour and restorative properties. A variant of the legend tells that the emperor tested the medical properties of various herbs on himself, some of them poisonous, and found tea to work as an antidote. Shen Nung is also mentioned in Lu Yu's famous early work on the subject, Cha Jing. A similar Chinese legend goes that the god of agriculture would chew the leaves, stems and roots of various plants to discover medicinal herbs. If he consumed a poisonous plant, he would chew tea leaves to counteract the poison.

A rather gruesome legend dates back to the Tang Dynasty. In the legend, Bodhi dharma, the founder of Chan Buddhism, accidentally fell asleep after meditating in front of a wall for nine years. He woke up in such disgust at his weakness that he cut off his own eyelids. They fell to the ground and took root, growing into tea bushes. Sometimes, another version of the story is told with Gautama Buddha in place of Bodhi dharma. Tea has played a significant role in Asian culture for centuries as a staple beverage, a curative, and a status symbol. It is not surprising, therefore, that theories of its origin are often religious or royal in nature.

(Source: History of tea - Wikipedia, the free encyclopedia; http://en.wikipedia.org/wiki/History_of_tea)

Chronology of Tea History

- ☆ In 2737 BC, the Emperor of China introduced tea to the world.
- ☆ In the year 350 AD , Kuo P'O of China referred tea in his book.
- ☆ In the year 780 AD, the first book on tea " Cha Ching (Tea Classics)" was published by a Chinese author, Lu Yu.
- ☆ Tea plants were transported to Japan in the 6th Century.
- ☆ From AD 1689 onwards, East India Company of England started importing tea from China.

Tea in India

- ☆ In 1823 AD, Robert Bruce and C.A.Bruce found wild tea growing in Sibsagar district of Assam.
- ☆ In 1834 AD, Lord William Bentinck, the Governor General of British ruled India appointed a committee to explore the possibility of growing tea in Assam.
- ☆ In 1839 AD , Dr. Campbell started an experimental nursery in Lebong, which is very near to Darjeeling town of West Bengal.
- ☆ In 1862 AD tea cultivation started in the Terai region of Bengal.
- ☆ In 1874 AD tea cultivation started in the Dooars region (Jalpaiguri district of Bengal)

Tea Research in India

- ☆ In the year 1891, Indian Tea Association and the Agricultural & Horticultural Society of Bengal jointly started tea research. M.Kelway Bamber was appointed for the purpose.
- ☆ In 1893, Dr. George Watt played a great role in tea research. He was basically an Entomologist.
- ☆ First tea research centre in India was established in 1904 at Heeleakah Tea Estate of Assam. Dr. Harold Mann was the Scientific Officer of that centre.
- ☆ This research centre was then shifted to Jorhat on the bank of the river Tocklai. The centre was named as Tocklai Experimental Station. This research centre is considered to the greatest and oldest tea research centre in the world.
- ☆ Tea Research Association was formed in the year 1964. TRA is the main organization which looks after the research work in tea in the entire North East India.

☆ UPASI Tea Research Institute in Coimbatore district of Tamil Nadu was established in 1926.

Systematic Position of Tea

Order : Theales

Family : Theaceae

Genus : *Camellia*

Species : *sinensis*

Botanically, tea is a tree. Tea tree is modified to a shrub by pruning with the following objectives :

(i) To keep it under pluckable height.

(ii) To induce lateral growth for producing more plucking points.

(iii) To keep the plant in vegetative form because the commercial product of tea comes from the leaves of the tea bushes and not fruit or seed.

(iv) To encourage more flush, more productivity.

Vegetative Characteristics of Tea

Characteristics	Average
Mean Leaf Angle (degrees)	85
Laminar Angle (degrees)	120
Inter nodal length (mm)	45
Individual Leaf Area (mm^2)	160
Leaf Area Index	6
Leaf Length / Breadth Ratio	2.4
Height(cm)	360
Girth at Collar (cm)	35
Branching habit Both orthotropic & plagiotropic Thickness of branches at 60 cm from ground level (cm)	3
Length of inter node between the second and third leaves from the apical bud of flush shoot (cm)	2
Angle between the 3rd leaf of the flush shoot and the inter node above (degrees)	50
Colour of mature leaf Light green to dark	green
Pubescence on the bud and abaxial side of the first leaf	Glabrous / densely pubescent (hairy)
Anthocyanin pigmentation in young leaves and petioles	Found in cambod variety
Dry weight of Flush (3 leaves and a bud) shoot (mg)	200

Tea Varieties

There are mainly three varieties of tea :

1. China Variety - *Camellia sinensis* (L) O.Kuntze.

It is a shrub with multistemming character. All the stems emerge from the soil directly. There is no distinct trunk. Leaves of China variety tea is very small and very high in Volatile Flavour Constituents (VFC) which imparts 'Muscatel Flavour'. Flavour Index is the ratio of VFC II and VFC I. In VFC IIs, Geraniol is the main component which emits rosy flavour . Apart from geraniol, it contains Linalool and its oxides. The leaves have orthotropic (erect) growth habit. Length of the leaves varies from 1.5 to 14 cm and width 1 to 2.5 cm. China variety is known for its flavour.

2. Assam Variety – *Camellia assamica* (Masters) Wight.

It is bigger in size than China variety . Shrub is vigorous in growth. Leaves are bigger than that of China variety. Length varies between 8 to 20 cm and width varies between 3 to 8 cm. The VFC content in Assam variety is less than that of China variety. Assam variety is known for its dark cup character with high concentration of polyphenols.

3. Cambod Variety – *Camellia assamica* ssp. *lasiocalyx* (Planchon ex Watt) Wight.

It has the features like Assam variety but has broadly elliptic (boat shaped) leaves with not too distinct marginal veins, often with pink anthocyanin pigmentation on the base of petiole particularly when young. Leaves are having orthotropic growth habit and pale green colour.

Anthocyanin pigmentation

2

CONDITIONS REQUIRED FOR GROWING TEA

Locality	Nearer to marketing place, transport and trading facility
Depth of Soil	Pervious up to 5-6 feet
Physical Properties of soil	Deep and porous.
Drainage	Good drainage with sufficient out fall
pH of Soil	4.5 to 5.5
Ambient Temperature	20 – 30 ° C (maximum 35 ° C)
Leaf Temperature	Min. 21 ° C (maximum 39 ° C)
Soil Temperature	Not less than 20 ° C; Optimum 25 ° C
Day length	Not less than 11 hours 15 minutes
Rainfall	1000 – 1400 mm of annual rainfall; 100-150 per month on an average
Relative Humidity	80-90 per cent

3

ECOPHYSIOLOGICAL ASPECTS OF TEA

Each plant is the product of its genetics and the total environment in which it is grown . The yield potential of any crop is set by the climate and its day – to – day variation which is known as weather. Climate sets limits for crop production. The main climatic elements in agriculture are temperature, moisture, sunlight and wind. All crops have certain natural threshold limits of the climatic elements beyond which they do not grow normally. Plant growth is dependent upon all the factors that make up the environment. No particular level of one factor should be referred to as the optimum for growth of a species without specifying at least the approximate levels or conditions of the other important components of the environment.

Climate determines where a crop is grown and the potential yield; the actual yield obtained depend on the prevailing weather. When nutrients are not limiting the important weather variables are solar radiation (sunshine), temperature, saturation deficit of the air and soil water availability. An analysis of how weather influences each of these processes can assist in (a) determining the potential yield in a given area (which is useful for planning purposes or when evaluating new sites available for commercial production) and (b) identifying other constraints to production which may be limiting yield and which could be correlated by appropriate management.

Characteristics of the plant or genotype, effect the way in which a crop will react to the environment. Many agricultural experiments are carried out that yield results of limited value because no consideration is given to important aspects of the environment, or no attempt is made to understand the interacting effect of cultural method on yield. The environment is of greatest importance in the life of the plant in as much as it is the cause of continuous response. Thus, the plant places itself in specific position in relation to gravity and incident light and it continuously adjusts itself to water supply and light intensity through stomatal movements, readjustment of its chloroplasts etc. The same plant may exhibit very different behaviour in different environment. The external factors which influence the rate and efficiency of photosynthesis are light, temperature, humidity and carbon dioxide. Photosynthesis is influenced by all the environmental factors *viz.* temperature, moisture supply, relative humidity of the air, age of the leaf etc. Acclimation of plants to different light environments have profound influences on the structure and function of photosynthetic apparatus. The main climatic variables influencing rates of shoot extension and yield of tea are temperature, rainfall, evaporation and the saturation deficit of the air and through their influence on plant and soil water deficit of the air and through their influence on plant and soil water deficits. It is difficult to specify the ideal or the average climate that tea requires. Photosynthetic efficiency is the primary component of dry matter productivity. It was reported that diseases and insect losses decrease U.S. yield below the genetic potential by only 4.1 and 2.6 per cent respectively. The remainder must be attributed to unfavourable physiochemical environments caused by weed competition, inappropriate soils, unfavourable climate and poor management practices. As far as climatic factors are concerned, according to the total indemnity made to U.S. farmers for crop losses – drought, excessive water and cold account for 71per cent.

Proper understanding of the interactions between environmental and physiological factors is essential for any scientific planning of tea cultivation. When climate plays an important role on physiology of plant, it is essential to study the behaviour of tea plant physiology under varying climatic conditions.

Ecophysiology of tea has several applications : it can be used to assist planners in assessing the yield potential of new (or existing) areas of tea and to estimate likely benefits from irrigation. It can also be used to assess the benefits of shade in different location and in the development of selection criteria for new clones, based on a knowledge of the base temperature for shoot growth. Further, it can be used to help to specify objective harvesting policies.

Source Sink Phenomenon

The permanent leaves below the plucking surface, which are not plucked are known as maintenance foliage. These maintenance foliage produce the assimilates which are supplied to other parts of a plant, which respire and grow actively with the help of those photosynthates. Hence, the maintenance foliage acts as the 'source' of photosynthesis and the tissues viz. bush frame, roots, flowers, fruits and the flush, which import those photosynthates are known as 'sink'. Partitioning of assimilate is governed by 'Source – Sink' relationship. It was reported that about 10per cent of photosynthates are harvested in the form of pluckable shoots and rest 90 per cent are diverted to unproductive parts like flowers, fruits, wood and roots. Mobilisation of photosynthates from the source to the sink depends on climatic conditions as well as abundance of sink. It was observed that in the month of December, 81per cent of the photosynthates was present in the maintenance foliage as against 64 per cent during June. It indicates that during summer because of active growth of shoots, the demand of photosynthates increases in the sinks, thereby inducing more mobilization from the source. Building of reserve during winter and mobilization of the same during spring has also been noticed. In a tea bush, growing shoots are the strongest sinks. Presence or absence of growing shoots on the bush surface determines the direction of food material flow.

A favourable balance between source and sink can be induced by adopting few field management practices *viz.* removal of dormant shoots, flower buds, addition of photosynthetically efficient maintenance leaf and spray of plant growth regulators.

Harvest Index of Tea

The harvest index of tea is very low compared to other field and horticultural crops. In tea, about 7% of the assimilates of a bush is harvested in the form of pluckable shoots whereas in other crops its about 20 to 30%. The harvest index of tea can be improved by harvesting more leaves, although this reduces quality and value of processed product. It has also been seen than harvest index varies in various clones. Hence, appropriate selection of tea clone can increase harvest index.

Crop Yield vs Photosynthesis

Very often the planters or the plant scientists are eager to select a clone or a seed stock having higher photosynthetic rate presuming that a plant variety having higher rate would produce more crop. The question is whether crop yield has any correlation with the quantum of net photosynthesis.

That a tea variety or clone having a photosynthetically favourable semi -erect leaf pose produced four times more crop than that having horizontal leaf pose. The number of shoots per unit area was six times higher in the plants bearing semi – erect leaves. This proves that semi – erect leaf pose intercepts more photon flux density which boosts the rate of photosynthesis and in turn augments crop yield. Usable productivity in tea is more directly linked to photosynthetic rate than in other perennial crops. Photosynthesis is the basis for dry matter production and in turn for the economic yield. Photosynthesis is the key to dry matter production and hence yield of economic organ. Thus increasing photosynthetic efficiency is the most important way of increasing productivity. But since tea crop is sink limited, further research into photosynthesis in tea is unlikely to contribute to increased crop yields. From these studies it could be inferred that the production of tea can be increased by increasing the sink capacity. A tea plant having a large population of actively developing shoots per unit area would be ideal for achieving high yield. It was observed that by doubling a plucking round, 38 per cent more yield could be achieved. But increase in plucking round has negative correlation with quality of tea. A solution to this problem could be to select clones which retain quality when shoots are allowed to grow larger. The yield of processed tea depends on the rate of shoot extension, the number of shoots per unit area, their fresh weight at harvest and the dry matter percentage. Out of these components, the number of shoots per unit area is mainly under genetic control . The size of the shoots depends on the harvesting policy of an estate. But the rate of shoot extension is mainly controlled by climatic factors.

Crop growth and production depend on the interaction of biological system which is the plant or more often a population of plants and the physical environment in which the plant grows. The total growth and production are in fact derived from the environment through the special mechanism and properties possessed by the biological system.

It was observed in China that temperature influenced the growth, development and yield of tea plant. In China, optimum temperature for most varieties was in the range 20 – 30 °C. Tea plants requires 1000 – 1400 mm of annual rainfall : 100 – 150 mm of rain per month except for the high temperature months. Relative humidity of 80 – 90% is favourable for tea plants, below which shoot growth was affected.

In Malawi, it was observed that shoot extension stopped below about 12.5 °C air temperature. Soil temperatures between 18 – 25 °C do not affect shoot extension. In Kenya it was observed that lower soil temperature affect growth and yield.

It was observed that maximum CO_2 uptake was at 30 – 35 °C. There was rapid fall beyond 37 °C and there was no net photosynthesis at 42 °C. Under natural conditions, temperature of fully exposed leaves was 2 °C to 12 °C higher than ambient temperature. In Japan, it was found that growth was most rapid at 30 °C and growth ceased below 12.5 °C. Tea leaf retain its peak photosynthetic ability for about six months though it continues functioning till about 18 months before it is shed. Photosynthesis is impaired at ambient temperature above 30 °C and seasonal dormancy is related to short day length and low temperature. In Kenya , it was observed that the minimum air temperature required to support shoot growth appears to be 13 – 14 °C with an optimum range of 18 – 30 °C. Day time maximum temperature in excess of 30° C and night temperature below about 14° C Probably lead to a reduction in the growth rate. The minimum leaf temperature necessary to initiate shoot extension is apparently 12 °C and at leaf temperature about 35 °C the rate of net photosynthesis falls off quickly. Long sunshine hours are probably essential for maximum yield. In most areas, 150 mm of rain each month i.e. 1800 mm of annual rainfall will ensure continuous crop production.

The main climatic variables influencing rates of shoot extension and yield of tea are temperature, the saturation deficit of the air and through their influence on plant and soil water deficits, rainfall and evaporation. The maintenance leaves below the plucking surface fix CO_2 photosynthetically and supply the photosynthates to other parts of the plant. All the mature leaves in the plants were photosynthetically active even under the cold weather conditions prevailing in December. He found that even the fifth leaf from the top was contributing photosynthates towards the growth of pluckable shoots at the top. He observed highest rate of photosynthesis in the topmost leaf of the maintenance foliage.

The topmost leaf in a canopy (maintenance canopy) contributed the highest proportion of photosynthates to the pluckable shoots. Pluckable 'two and a bud' also photosynthesizes, but to a much lesser extent than the mature leaves. Maintenance foliage in tea is the main source of carbohydrate.

Experimentally it has been proved that maintenance foliage carry out photosynthesis throughout the year. Therefore, it is not the stoppage of photosynthesis which makes the bushes go dormant. Inter-flush dormancy is caused by low availability of moisture and various other factors whereas winter dormancy is caused by the interaction of some environmental factors like short day length and low temperature.

Light intensity and CO_2 concentration are the primary external parameters which control photosynthesis. Temperature is a secondary factor. The effect of temperature on photosynthesis varies with light intensity, and the carbon dioxide supply, Both these factors control the photochemical reaction, which are generally independent of temperature. With low light intensity and adequate CO_2 or with low CO_2 and adequate light, the light reactions will be limiting. Reed (1946), however, inferred that photosynthesis is influenced by all the environmental factors *viz.*, temperature, moisture supply , RH of the air, age of the leaf. CO_2 concentration affects many plant processes, mainly through its direct effect on photosynthesis and stomatal physiology.

Effect of Temperature

Temperature is a major factor for growth. In temperate and subtropical regions of the earth, temperature constitutes a major limiting factor in crop production ; in particular the low temperature which occurs in winter and early spring. On the other hand, in the humid tropical zone, it is very unlikely that temperature alone is ever a serious limiting factor in crop production. There are, however, special situation within tropics in which temperature becomes important namely the cultivation in high altitude areas. Even small variations (approximately 1 °C) in air temperature can have pronounced effect on shoot extension rates. In general, mean minimum temperatures below 13 °C are likely to bring about damage to the foliage and retardation of growth : Mean maximum temperature above 30 °C are likely to be accompanied by humidity so low that a similar cessation of active development is inevitable because plants are not able to regulate their temperature effectively, their distribution, both on a worldwide and smaller ecological scale is strongly influenced by ambient temperature. Every plant has a optimal temperature for growth, and upper and lower limits, beyond which all growth stops. Because temperature effects metabolism, it also influence growth. It is interesting that temperature of tea is not very location specific, but it may vary genetically between cultivars under comparable conditions of growing. Tea exhibits a wide range of temperature tolerance and is cultivated from the humid equatorial regions to subtropical and temperature latitudes. High altitudes in the tropics and subtropics are also commonly cultivated with tea. Effect of low temperature on plant apply to temperate species as well as tropical species. Translocation of assimilates is also reported to be reduced by low temperature and this may be expected to affect photosynthesis through the accumulation of photosynthates at the primary reaction centers. A further general effect of low temperature in both tropical and temperate plants appears to be a change in the distribution of photosynthate between shoots and roots, the roots receiving proportionately more assimilates. Photosynthetic response to temperature was mainly due to change in mesophyll activity rather than stomatal conductance. Temperature limits for photosynthesis in the intact cell are given on the one hand by the damage of essential structure due to disruptive crystal formation and on the other hand by heat denaturation of enzymes and structural proteins. However, the photosynthetic reaction is not strongly affected by the ambient temperature, as long as lethally high or low temperatures are not encountered.

Effect of Light

Light has three major effects on photosynthesis. First, it provides the energy used in the production of ATP and NADPH. Second, light promotes the activation of key enzymes in the photosynthetic apparatus and stimulates stomatal opening. Third, light acts through photoreceptors such as phytochrome to modulate the development of leaves and the photosynthetic apparatus light promotes enzyme activation largely through its effect on electron system. Important light activated enzymes in photosynthesis include rubisco, stromal fructose – 1,6 biphosphatase, NADP – triose phosphate dehydrogenase, ribulose – 5 phosphate kinase and sedoheptulose biphosphate.

The radiation reaching the surface of the earth from the sun has a maximum intensity of about 1.6 calorie per cm^2 per minute, corresponding to a visible irradiance or about 1,00,000 lux or meter – candles. The intensity, and to a lesser extent, the spectral composition of solar radiation varies, of course, with time of day, weather , season and geographical position. Plants must made the best use they can of this mixed and variable

radiation for the energy they need for photosynthesis. Variations in light intensity and duration of illumination affect photosynthesis directly. The vast majority of this short wave radiation is interrupted by leaves in the top 0.3 m of the canopy, almost regardless of foliage geometry below 0.1 m with only about 5% reaching the ground. Only a very small proportion of the net available energy is used in the process of photosynthesis, the vast majority is dissipated as latent heat (through the process of evaporation) and as sensible heat (heating the surrounding air). Transpiration increases with increase in photosynthetically active radiation. Sunlight may fall on leaf either directly, after reflection by molecules in the air (sky – light) or after reflection from clouds (cloud light) of other surrounding object. The amount of solar radiation incident on the earth atmosphere , the so – called "solar constant" is about 8 – 4 J per minute and of this about 50% reaches the earth surface at sea level on a cloudless day when the sun is directly overhead light intensity reached a maximum of about 10,000 lux under these conditions. The amount of radiation observed by a leaf depends on the wavelength and on its spectral characteristics. About 44 per cent of the solar energy incident on the earth surface is useful for photosynthesis. The maximum light energy that a tea leaf can use for photosynthesis varies from clone to clone even under optimal conditions . The study of growth and yield of a plant requires the Knowledge of the photosynthetically active radiation (PAR). PAR is defined as the part of the solar spectrum between the wavelength 400 – 700 n.m. (0.4 – 0.7 um).

Light affects crop growth and production both through its use in photosynthesis and through photoperiodic reactions. Two aspects of photosynthetic light are of interest in studying crop production; the total amount of incoming light which is suitable for photosynthesis and the amount of light which is available for or can be utilized by the crop. Light as one of the most important parameters for photosynthesis because of its vital role as energy producer.

Reduction in photosynthesis under low light could be attributed to high stomatal resistance to CO_2 exchange. Under low light; reduction in photosynthesis occurred, on the other hand, there is abundant evidence that photosynthesis is inhibited due to high light intensity. Exposure of plants to high light intensities can reduce photosynthesis.

Sub less than 900 µmol m^{-2} s^{-1}(micro mole per metre square per second) and supra optimal (greater than 1200 µmol m^{-2} s^{-1}) levels of PAR inhibited photosynthesis significantly.

Effect of Water Stress

1. There is a general reduction in size of the plants subject to prolonged water stress.
   a) Reduced cell turgor is the most important reason for reduced plant size.
   b) Changes in structure of proteins caused by decrease in water potential effect enzyme activity.
   c) Osmotic potential is important since the protoplasm must be in osmotic equilibrium with the sources. Water stress decreases O.P. Small decrease in osmotic potential may be accompanied by a large decrease in turgor pressure and water potential.
2. Photosynthesis is reduced by closure of stomata.
3. Reduction in leaf area, cell size and intercellular volume.
4. Reduces synthesis of DNA, RNA and cell wall materials. DNA content decreases from 80 – 40 per cent of the control leaf values with a decrease in leaf water potential from – 2 bars to – 8 bars.
5. Dehydration of protoplasm.
6. Decrease in starch content.
7. Decrease in soluble protein.
8. Inhibits auxin formation.
9. Cytokinin activity reduced.
10. Translocation reduces.
11. Proline (Pyrolidine – 2- Carboxsylic acid) increased.

Moisture is one of the most important parameters for physiological activity and growth of plants. It had also been reported by various workers that low moisture is a limiting factor for physiological activity.

Reduction of the photosynthetic rate as a result of moisture stress is well documented . Reduction in water content usually reduced the rate of photosynthesis both because dehydrated protoplasm is less effective and because stomatal closure reduces the supply of CO_2. Decrease in photosynthesis at high values of stress may be due to stomatal closure. Inhibition of photosynthesis by water stress may be attributed to both stomatal and non- stomatal factors depending on particular environmental conditions. Drought – induced reduction in translocation was proportional to the decline in photosynthesis because the amount of water actually needed for the photosynthetic process would be very little in comparison with the amount needed to maintain the living plant. Therefore, since long a deficiency of the supply of water actually needed for the photosynthesis in a direct manner, the indirect effects of a shortage of water could impair vital processes of the biological mechanism retarding photosynthesis. Shortage of soil water or atmospheric water stress on the plant will affect the efficiency of the photosynthetic reaction.

Humidity is of importance in tea physiology primarily because of its influence in determining the loss of moisture by evapo transpiration. The success of tea in high altitude sites partly appears to be related to the crops requirement for high moisture levels . Water is an essential constituent of active protoplasm and is a reagent or reactant in various physiological processes including photosynthesis. Only 1 per cent of the available liquid water taken by plants is actually involved in metabolic activities.

Tea bushes are benefited by high atmospheric humidity. He added that low humidity affects the physiological activities adversely by influencing the cell sap concentration. Water stress in the plant directly reduces the rate of photosynthesis because the dehydrated protoplasm has a lower photosynthetic capacity.

It has been reported that irrigation plays an important role in combating moisture stress. Strategic sprinkler water helps more than furrow or ground irrigation. Internal water stress can be reduced by ground irrigation but it does not have any lasting effect on temperature and vapour pressure of the atomosphere except for the short periods when water is actually being applied overhead. Intermittent sprinkling of water was found to reduce temperature and increase humidity around tea bushes in Russia. An experiment conducted in Malawi showed that misting the tea in hot dry months removed adverse effects of dry air and allowed shoots to grow at the same rate as shoot growing at the same mean effective daily temperature in the rainy season.

Water stress can reduce photosynthesis by reduction of leaf area and closure of stomata. It affects practically every physiology and biochemistry. High atmospheric stress is expected to reduce photosynthesis probably causing rapid evaporation from guard cell and causing stomata to close. Decreasing water potential in leaves is known to reduce photosynthesis. Water stress reduce photosynthesis mainly because of hydro active closing of stomata, bringing about a reduced CO_2 sypply; water stress in cytoplasmic ultrastructure affecting enzyme activity and dehydration of cuticle, epidermal walls and cell membranes reducing their avidity for and permeability to CO_2.

Reduction in water content usually reduce the rate of photosynthesis both because dehydrated protoplasm is less effective and because stomatal closure reduces the supply of CO_2. Photosynthesis and water economy of terrestrial plants are so intimately interrelated that water supply is normally one of the most important factors determining the amount of photosynthesis carried out. The decrease in photosynthesis at high values of stress may be due to stomatal closure, which has been shown to limit photosynthesis in water stressed leaves. Water stress reduced photosynthesis due to impaired stomatal conductivity. Higher stomatal resistance could also affect gaseous exchange and consequently, the rate of photosynthesis. Under low light, reduction in photosynthesis occurred. Stomatal resistance is the principal limitation to photosynthetic CO_2. Since stomatal opening is more dependent on water balance than on light, photosynthesis may be limited severally during the period of water stress despite the fact that light and CO_2 supply are optimum. Decrease in net photosynthesis with increased water stress. Temperature limits photosynthesis by the neat denaturation of enzymes or structural proteins. According to him the external factors which influence the rate and efficiency of photosynthesis are light, humidity and CO_2. Drought reduction in translocation was proportional to the decline in photosynthesis because the amount of water actually needed for the photosynthetic process would be very little in comparison with the amount needed to maintain the living plant. Therefore, since a long deficiency of supply of water became inhibitory to photosynthesis in a direct manner, the indirect effect of a shortage of water could impair other vital processes of the biological mechanism retarding photosynthesis. It has been noticed that net photosynthesis was lower in developing than in mature and old leaves. Low temperature and day length apparently interact including dormancy but their respective roles are not yet clear. Several scientists have attempted to select plant with high stomatal resistances increases under water stress condition but the conductance reduced. Stomatal resistance increased under drought stress. It is known that moisture stress increases stomatal resistance . With slowly imposed water stress, it may be the effect of water deficit in reducing net photosynthesis that reduces stomatal conductance. Stomatal conductance decreased significantly due to stress in all genotypes. Increased temperature of the stressed plant is due to stomatal closure. Leaf extension rate is a linear function of temperature. High temperature exerts two opposing effects upon the stomata – (i) at a given light intensity, the minimum intercellular CO_2 concentration, increases with temperature (ii) if excessive accumulation of CO_2 in the intercellular spaces does not occur, raising the temperature causes opening. Leaf to air temperature difference was significantly higher under stressed condition.

Studies showed that transpiration rate is directly proportional to plant production. Decrease in transpiration rate is observed with increase in stress. Transpiration is controlled by the atmospheric factors which determine evaporation as well as by plant factors such as stage of plant growth, leaf area, leaf temperatures and soil moisture. He also reported that transpiration varies greatly with season. Transpiration ratio under stress are lower for resistance than for sensitive genotype.

Soil moisture deficit also reduces the transpiration rate. In China type of clones the transpiration was lower at the higher soil moisture level, but its depression was minimum at low soil moisture status . Soil moisture stress decreases transpiration rate. Wind usually causes increased transpiration because it removes water vapour from the vicinity of transpiring surfaces and produces a steeper vapour pressure gradient from plant tissue to air, but it also tends to cool leaves, thereby decreasing the steepness of the vapour pressure gradient.

Effect of Wind

Wind turbulence could be useful in reducing high temperature which otherwise would adversely affect photosynthesis. However, direct effect of wind speed on physiology of growth and productivity of tea are not known. Air movement is an essential feature for plant growth. Rather low velocities of air movement, however, would be sufficient to ensure adequate air renewal at the leaf surface. In relatively still air, a linear relationship between net photosynthesis and the wind velocity can be established. Wind increases transpiration so that water deficits are liable to occur sooner and stomatal closure may then reduce photosynthesis. Secondly, renewal of air at the leaf surface will maintain the CO_2 concentration around the leaf at normal levels under conditions of rapid CO_2 uptake. With limiting moisture, wind may be expected to reduce photosynthesis, on the other hand, with a favourable moisture balance for the plant, wind is likely to increase photosynthesis.

Role of Water Potential

Decreasing water potential in leaves is known to reduce photosynthesis. Photosynthesis decreases with low leaf water potential. Drought tolerant clones consistently exhibited higher water potential. Water stress lowers the water potential values, but water potential of cold stressed plants was not altered at all. Leaf water potential was lower when photosynthesis was reduced. There is no significant difference in water potential between the ages of plants. It has been established for a long time that low water potential causes metabolic imbalances within plants, cell elongation being one of the most sensitive portion . During the very winter months, value of water potential decreased than in the wet summer months. Clones of China origin exhibited higher water potential compared to Assam clones.

Role of Stomatal Functions

Photosynthesis controls stomatal conductance. With slowly imposed water stress, it may be the effect of water deficit in reducing net photosynthesis, that reduces stomatal conductance. Leaf photosynthesis increased with increasing leaf conductance. Reduction in stomatal conductance due to water stress. Higher stomatal resistance could affect gaseous exchange and consequently the rate of photosynthesis. Stomatal resistance is the principal limitation to photosynthetic CO_2 Photosynthetic resonse to temperature was mainly due to change in mesophyll activity rather than stomatal control. High stomatal resistance in plants would be advantageous in water economy and resistance. Decrease of photosynthesis is observed with decrease in stomatal conductance. Stomatal resistance is increased under water stress condition. Stomatal resistance increases and photosynthetic rate decreases as a result of moisture stress.

Role of Leaf Temperature

Leaf temperature directly affects plant metabolic activities. Leaf temperature above 35° C is too critical for photosynthesis. Transpiration plays very important role in the hydraulic cycle of crops, while transpiration is an active plant response to climatic factors, leaf temperature is a passive outcome of the heat and mass exchange. Leaf temperature directly affects plant metabolic activities. Temperature regulates leaf biochemical process rates. Leaf temperature deviates from ambient air temperature in a manner dependent on solar radiation level and ambient vapour pressure deficit. Leaf temperature necessary to initiate shoot extension is apparently 12 °C and at leaf temperature above 35 °C, the rate of photosynthesis falls off quickly, The genotypes with the warmest leaves under water stress conditions are most resistant to drought than cool leaved genotypes.

4

PROPAGATION OF TEA

There are many types of propagation methods in tea but clonal propagation is the most popular and acceptable method of tea propagation. Breeding of Tea through the techniques of tissue culture, genetic engineering, mutation breeding , polyploidy are still in the experimental stage.

Different Types of Propagation Methods and Breeding of Tea

- Vegetative propagation
- Seed propagation
- Grafting
- Composite plant
- Hybridization
- Genetic engineering
- Tissue culture
- Mutation breeding
- Polyploidy
- Recombination

Vegetative Propagation

A form of asexual reproduction in plants, in which multicellular structures become detached from the parent plant and develop into new individuals that are genetically identical to the parent plant. Vegetative propagation in tea is done by clonal cuttings. Cuttings are taken from scions of a desired tea clone. Cuttings may be single nodal or binodal. In case of single nodal cutting, a leaf with an axillary bud and 2.5 cm stem is planted. In case of binodal cuttings, two leaves with axillary buds are taken. Binodal cuttings are popular in African countries.

Seed Propagation

When plant is produced through seed it is called seed propagation. Generally biclonal seeds are used in tea gardens. Seeds are procured from the tree plantation of two different kinds of clones with different characters. The method of natural cross fertilization is adopted in seed production. Two kinds of clones are planted in alternate sequence so that one plant of a particular clone is surrounded by another clone on four sides, thus increasing the probability of cross fertilization.

Tea Seeds

The seeds are collected from the ground and packed in a box with charcoal for conserving moisture so that embryo does not get dried up. The seeds are then subjected to a sinker floater test. Seeds are dipped in water. The seeds which float because of dried embryo are rejected and sunken seeds are placed on a moist sand bed and covered with thin layer of thatch. The cracked seeds are transferred to the seed nursery and sown in the soil bed directly or in a polythene sleeve filled with soil with the cracked side downwards.

Hybridization

This is a method to amalgamate the desirable characters of two different clones having different traits and produce a plant with higher vigour, productivity, quality and pest resistant or drought resistant characteristics by pollen transfer method (Cross breeding). Pollen grains from the flower of a desired clone are dusted with a brush or gently touched on the stigma of the flower of another desired clone which remains receptive for 24 hours before anthesis up to two days after pollination.

- The flower buds soon after pollination are bagged with butter paper or muslin cloth bags for 48 hours to avoid out crossing.
- Then fertilization takes place and fruits attain hard seed coat.
- Each capsule generally contains one to three, sometimes four seeds of 1.5 to 2.0 cm diameter.
- The seeds have a hard testa outside and the embryo is covered in between two large cotyledons.

Planting of seeds

- After collection, seeds are brought to a central shed passed through a rotary type sifter to eliminate the very small seeds.
- After elimination the very small seeds, the remaining seeds are transferred to a tank which is filled with water and allowed to soak 2-3 hours.
- The tank should be deep enough to allow at least 15 cm of water to remain above the level of the seeds that sink.
- The sinker floater test is conducted and sunken seeds are taken out of water and spread on a concrete floor, tarpaulin or a plastic sheet until the adhering water evaporates.
- The seeds that float on the surface and those that remain suspended in water without sinking to the bottom are removed. The seeds floats because of the drying of embryo.
- Then the suitable seeds are planted in rows as soon after harvest as possible at depth of about 1.5 cm pointing downward or parallel to the ground surface.
- The distance between seeds should be about 20 cm.
- The seeds are also planted in polyethelene sleeves.
- The nursery management is almost same as vegetative propagation.

Grafting

- A technique through which vegetative part of a plant can be mechanically fixed on another plant which grows like a single plant after successful union.
- Grafted part is called SCION and the basal part on which it is grafted is called ROOT STOCK or STOCK.
- Scion provides the canopy of the bush maintaining all characters like phenotypic, genotypic, physiological and biochemical.
- 4 Stock takes the responsibility of nourishing Scion derived canopy by absorbing water and nutrients from the soil through strong root system and transporting them to the utilization.
- The increased interest in grafting in tea estates has focussed attention on the problem of graft failure, which might be due to "Compatibility barrier" between the stock and the scion. Some chemical compound like levels of polyphenols and catechins play an important role in the success of grafting.

  Suitable time for grafting is November to February.

Methods of Grafting

(a) Cleft grafting

(b) Rind grafting

(c) Bud grafting

Cleft Grafting

- A cleft is done on the clean cut stock branch at the centre with the help of a sharp pruning knife, for widening the cleft a bamboo wedge is inserted at the centre.
- In the scion a slashing cut is done in the internode,0.5 - 1.0 cm below the lower leaf, from both sides to form a tapering end.
- Prepared scion is inserted into the cleft made on the stock branch in such a way so that cambium layers of both scion and stock come in close contact and remain in one line.
- After removing the bamboo wedge the grafting union is covered by coir fibre and the entire graft is covered with a polythene bag and the bag is tied from below and kept under overhead shade.

Rind Grafting

- Rind grafting is Similar to cleft grafting except Scion is fixed on the stock branch on the side with out making the cleft.
- Stock branch is cut to convenient height by peeling the bark 3.5-5 cm length in one side, keeping the lower side intact.
- Edges of the cut bark is slightly lifted to fix the L shaped cut scion of 2.5-3.5 cm long from one side.
- Then it is tied up and kept under shade.

Bud Grafting

- This method is used in odd bushes where cleft or rind grafting cannot gives the result.
- Firstly an inverted T ,1.5x3 cm long cut is made in the bark of a Strong selected stock shoot.
- A bud from the bud wood with about 2 cm of bark is cut as scion.
- Scion or bud is inserted in the inverted T cut on the stock Branch.
- Gently it is tied up with Polythene strip keeping the axillary bud outside of wrapping.
- Any growth below the graft must be removed till bud becomes shoot.

Post Grafting Care

- The graft and the overhead shade should be observed continuously and no sun flecks should be allowed to reach the grafted scions.
- Watering should be done time to time.
- When the axillary buds of the scion grow and touch to the Polythene bag it should be removed.
- When the axillary buds of the scions produce 2-3 flushes, shade should be gradually thinned out and finally removed.

Composite Plant

This is a grafting method at nursery stages. The scion of a suitable clone having high yield potential, high quality, pest and drought resistant qualities is grafted on the stock of another suitable clone having good rooting system. The scion maker and graft maker machines are used to cut tapering end of the scion and groove in the stock. Since grafting is completely a mechanical association and there is no interchange of genetic material, the bottom always will be a good rooter while the upper portion will produce either a high yielder or high quality or pest resistant or drought resistant character.

A grafting involving nodal segments from primaries of two different but compatible clones leads to the formation of a composite plant. The top vegetative part –the scion of the mechanically fused plat produces tea similar to those superior clones which otherwise do not grow well in such areas. The stock clone with its good rooting ability supports the growth and development of the top vegetative part.

Collection of Primaries

Cut the primaries from stock and scion mother bushes separately during morning hours just prior to grafting .Dip the cut ends into buckets containing clean water. Bring the buckets to the nursery site.

Preparation of scion and stock

- ☆ Semi hard regions of the primaries should be used.
- ☆ The stock may be about 6-7cm long having two internodes and a healthy mother leaf at the mid nodal position.
- ☆ Scion may also be about 6-7 cm long having a mother leaf at the top nodal.

Nursery Grafting Method

- ☆ Cut open the top end of the stock with a sharp knife .
- ☆ The basal region of the scion should be cut from both the sides in the shape of a wedge.
- ☆ The length of the tapered portion should be equal to the depth of the cleft of stock .
- ☆ After insertion of the scion , tie up the graft portion with a 1.0 cm wide polythene stripe .
- ☆ Plant the composite cuttings in a sleeve (15.0cm lay flat and 22.5cm -25 cm in length).
- ☆ Cover with polythene bag tied around the sleeve with a rubber band .
- ☆ Place the sleeves in the nursery bed covered by overhead shade .
- ☆ Best done during spring and autumn.

Post Grafting Care

- Scions and stocks in nursery grafting and composite plants should not be treated with fungicides or hormones etc.
- When shoots coming out should always be removed .
- Maintain the soil moisture .
- Weeding is done.

Clones for Nursery Grafting

- Nursery grafting is done between two compatible clones. For assessing compatibility ; graft different clones in a smaller scale. Assess callusing and fusion within 6-8 weeks. If success is 50 per cent and above the clones can be considered as compatible.
- Clones suitable as SCION- TV1, TV2, TV14, TV17, TV21, S3A1, S3A3 and T3A3.
- Clones suitable as STOCK – TV9, TV18, TV20, TV21, TV23, TV25 , TV26, TV27, TV28, TV29and TV30.

Tissue Culture / Single Cell Culture / Protoplast Culture/Embryo Culture

Plant tissue culture is a collection of techniques used to maintain or grow plant cells, tissues or organs under sterile conditions on a nutrient culture medium of known composition. Plant tissue culture is used to produce clones of a plant in a method known as micro propagation. In tea industry, tissue culture is not very popular.

Mutation Breeding

- Mutation breeding is the process of exposing seeds to chemicals or radiation in order to generate mutants with desirable traits to be bred with other cultivars. Plants created using mutagenesis are sometimes called mutagenic plants or mutagenic seed.
- The technique has been greatly potential for breeding simple inherited characters like cup quality and resistance against disease, insect pests and drought.
- Cotyledon, embryo and even the endosperm can be irradiated and if cultured *in vitro* to regenerate plantlets, may open up an encouraging area of studies in genetics and breeding of tea.
- There are several reports on production of aneuploidy by mutation which have wide scope in tea breeding.

Polyploidy

- Polyploid cells and organisms are those containing more than two paired (homologous) sets of chromosomes. Polyploidy refers to a numerical change in a whole set of chromosomes.

Polyploidy in Tea plant

Tea plants under cultivation are mostly diploids with two set of chromosomes. Diploid chromosome number of tea is 30.

- Cup quality of most of the triploids was found to be either at par with that of the diploid clones or even better.
- Tea polyploids have clearly demonstrated the superiority of triploids and tetraploids over domesticated diploids in morphological and agronomic characteristics.
- All the triploids had higher values for various leaf characters than their diploid parents.
- So, polyploidy is an essential part of breeding in tea plant.

Clonal Selection

This selection method is conducted in an open field in a tea section. Plants with suitable characters are selected and observed. Following methods are followed in case of clonal selection.

Morphological Characters

(1) The canopy should be strong and devoid of any disease damage symptoms.

(2) Number of plucking points should be more. The number of plucking points can be counted by putting a quadrate of 50 cm x 50 cm dimension on top of the canopy.

(3) The bush should be free from pest infestation or at least should have low attack of pests.

(4) The growth habit of the branches should be orthotropic i.e. erect habit. This facilitates more penetration of solar energy (photosynthetic photon flux density) to the lower leaves and more leaves can photosynthesize and produce more glucose. In case of plagiotropic growth habit, the upper leaves obstructs the incidence of sunlight.

(5) The colour of the maintenance foliage should be pale green and not dark green. Dark green leaves indicate higher concentration of chlorophyll which in turn reduces the quality of tea.

(6) The flush of leaves should be high which would indicate that the bush is high yielding.

(7) The tips (apical buds) should have more pubescent hair. Abundance of pubescent hair is an indication of high quality.

(8) The bushes with wilting symptoms during drought should not be selected.

(9) The bushes badly affected by water logging in a low lying area should not be selected.

Physiological Characters

With the help of a Photosynthesis system, the following physiological characters are studied:

(i) Net Photosynthesis (Pn) (µmol $m^{-2}s^{-1}$) - Higher rate of net photosynthesis will ensure higher growth rate, higher biomass production and in turn higher productivity.

(ii) Transpiration (E) (mol $m^{-2}s^{-1}$)- Low transpiration rate during drought (water stress period) and higher transpiration rate during rainy season are desirable.

(iii) Stomatal Conductance (gs) (cm s^{-1})- Lower stomatal conductance during drought and higher stomatal conductance during rainy season are desirable.

(iv) Stomatal Resistance (rs) (s cm^{-1}) - Lower stomatal resistance during rainy season and higher stomatal resistance during drought are desirable.

(v) Leaf Temperature (T_L) (°C) – optimum is 21 °C.

(vi) Intercellular CO_2 concentration (ppm)- Higher Intercellular Concentration will ensure more photosynthesis.

For the afore stated studies, the leaf discs of the first leaf of the maintenance canopy should be used.

With the help of a Dew Point Micro voltmeter the following parameter are studied:

Leaf water potential (Ψ) (-bar)- Higher rate of leaf water potential imparts drought tolerant character.

Stomatal Dimension

Stomatal frequency per unit area of leaf (the first leaf of the maintenance canopy) and the size of the stomatal aperture should be studied. A leaf area meter should be used to measure the maximum width, minimum width, area and length of the leaves.

For this study the nail polish method should be used and the measurement of the stomatal dimension should be done by using ocular and stage micrometer.

If the stomatal frequency per unit area is more and the aperture is bigger then it will indicate that it is not suitable for a droughty area. The total pore area should be less.

Biochemical Parameters

The first leaf of the maintenance canopy should be taken for study of the biochemical parameters.

(i) Total free Amino Acid (µ gm / gm of leaf) – These are the amino acids which do not take part in protein synthesis. Higher content of total free amino acids is desirable. Moore & Stein (1948) method can be applied for the estimation of total free amino acid.

(ii) Proline (µ g / g of leaf) – Higher proline content imparts drought tolerance to a tea plant. Hence, for a droughty area, a plant with higher proline content should be selected. Bates *et al* (1973) method can be applied for the estimation of proline.

(iii) Ascorbic acid (mg / gm of leaf)- Ascorbic acid is Vitamin C and it provides disease preventing character to a tea plant. Harris & Ray (1935) method can be applied for the estimation of ascorbic acid.

(iv) Chlorophyll content (mg / gm of leaf)- Higher chlorophyll content reduces the quality of tea. Arnon (1959) method can be applied for the estimation of chlorophyll.

(v) Epicuticular Wax (mg / gm of leaf) – Wax content reflects sunlight and reduces transpiration loss. Hence higher concentration of epicuticular wax induces drought tolerance. Silva Fernandes *et al* (1964) method can be applied for the estimation of epicuticular wax.

Rooting Trial

Let the shoots grow by leaving it unplucked. Take cuttings from the scions and plant in nursery sleeves. After about 3 months, uproot the plants carefully and study the proliferation rate and health of the roots. Selection can be made on the basis of rooting performance.

Multilocational Trial

On the basis of above selection processes, select 10 good bushes and these 10 plants should be put to Multilocational trial (hilly area, plains, droughty area, waterlogged area, sandy soil etc) for final selection. Cuttings from those 10 bushes should be taken , grown in a nursery and the plants may be planted in different locations and again repeat the entire process as described to find the suitable clones for releasing to the industry.

5

SOIL SCIENCE AND NUTRIENTS

Soil is defined as "The uppermost loose layer of the earth suitable for plant growth."

Components of Soil

Mineral Matter (45%)

Organic Matter (5%)

Soil Water (25%)

Soil Air (25%)

Difference Between Top Soil and Sub Soil

Top Soil (0-22.5 cm from top surface)	Sub Soil (below 22.5 cm depth)
Completely weathered	Not completely weathered
Soil particles are fine	Particles are coarse
Aeration is more	Poor aeration
More number of micro-organism	Number is less
Higher organic matter content	Lower organic matter content
It is dark in colour	It is light in colour
High in Nitrogen content	High in potassium content
It is fertile	Sub soil is less fertile

Physical Properties of Soil

Physical properties of a soil influences plant growth. Following are the important physical properties of soils:

Soil texture, Soil structure, Soil density, Soil porosity, Soil consistency, Soil colour, and Soil temperature.

Soil Separates

According to the size, the soil particles are grouped into gravels, sands, silts and clay which are termed as Soil Separates.

Soil Separates	Diameter Range(mm)
Coarse Sand	2.00 – 0.20
Fine Sand	0.20 – 0.02
Silt	0.02 – 0.002
Clay	Below 0.002

Mechanical Analysis

The process of determining the amounts of individual soil separates below 2 millimetres is called Mechanical Analysis. There are several methods of mechanical analysis viz, Sieve Method, Sedimentation Method, Pipette Method, Hydrometer Method etc. Pipette and Sedimentation methods are universally used for mechanical analysis. Mechanical Analysis helps in deciding the textural classes' name – that is sand, sandy loam, loam, sandy clay loam, sandy clay etc.

Soil Texture

Characteristics	Sand	Silt	Clay
Size	2.0 – 0.02 mm	0.02 – 0.002 mm	Below 0.002 mm
Visibility	Visible by naked eyes	Visible by microscope	Visible by ultra microscope
Water holding capacity	Low	Medium	High
Movement of air and water	Very rapid	Moderate	Slow
Plasticity	Very low	Moderate	Very high
Feel on rubbing between thumb and fingers	Gritty	Like talcum powder	Feels very plastic and sticky in wet condition and become hard under dry condition
Fertility	Very low	Moderate	High
Minerals	Quartz dominates	Feldspar, mica, hematite quartz etc	Kaolinite, illite etc.
Chemical activity	Inactive	Slightly active	Very active

Textural Class

Textural group	Sand %	Silt %	Clay %
Sand	80-100	0-20	0-20
Sandy loam	50-80	0-50	0-20
Loam	30-50	30-50	0-20
Silt loam	0-50	50-100	0-20
Sandy clay loam	50-80	0-30	20-30
Silty clay loam	0-30	50-80	20-30
Clay loam	20-50	20-50	20-30
Sandy Clay	50-70	0-20	30-50
Silty Clay	0-20	50-70	30-50
Clay	0-50	0-50	30-100

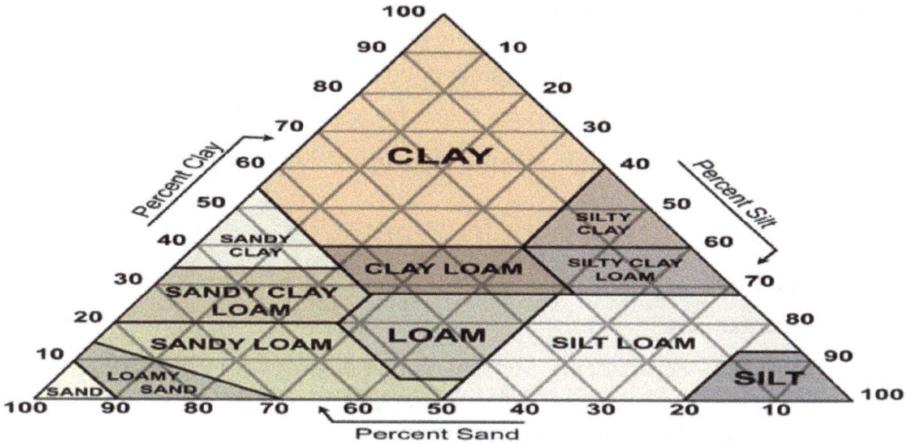

Soil Structure

There are four principal forms of structure:

(1) Plate-like – This type of soil structure is mostly noticeable in the surface layers of virgin soils but may be present in the sub soil.

(2) Prism-like – This type of soil structure is commonly occurring in sub soil horizons.

(3) Block – like - This type of soil structure is mostly noticeable in the sub soil horizons and characteristics have much to do with soil drainage, aeration and root penetration.

(4) Sphere- like – All rounded aggregates may be placed in this category. These rounded complexes usually lie loosely and separately.

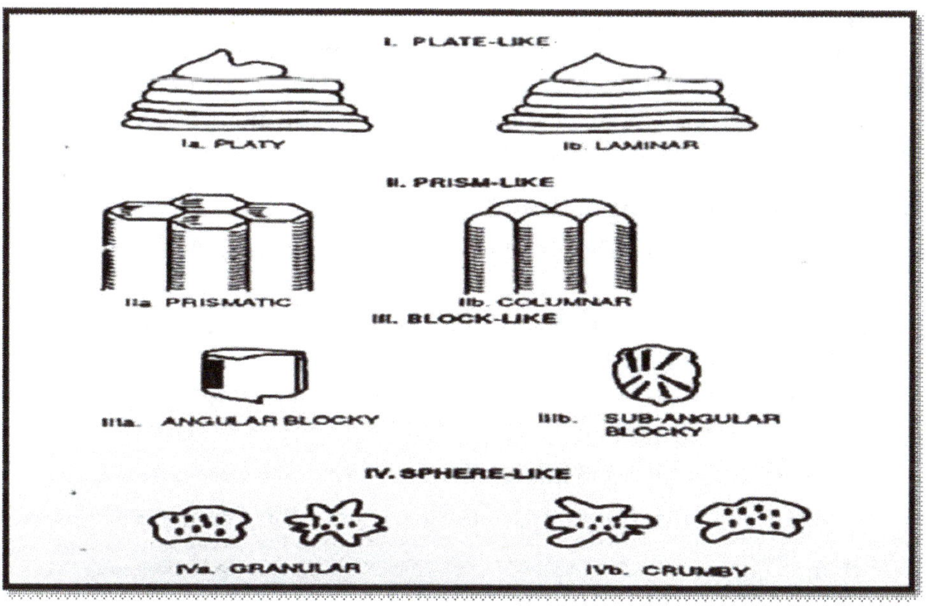

Factors Affecting Soil Structure

The development of structure in soil depends on the following factors:

(1) Climate

(2) Organic matter – It improves the structure of a sandy soil as well as of a clay soil.

(3) Tillage

(4) Plant Root

(5) Soil organism

(6) Fertilizers – Fertilizer like Sodium Nitrate destroys granulation by reducing the stability of aggregates. Few fertilizers, for examples, Calcium Ammonium Nitrate help in development of good structure.

Soil Density

Particle density - The weight per unit volume of the solid portion of soil is called particle density. Particle density of a normal soil is 2.65 g/ c.c.

Bulk Density – The oven dry weight of a unit volume of soil inclusive of pore spaces is called bulk density. The bulk density of soil is always smaller than its particle density. The bulk density of sandy soil is about 1.6 g/c.c.

Porosity of Soil

Soil porosity is the percentage pore space. It can be calculated by the formula:

Porosity (% pore space) = 100 – (bulk density / particle density) X 100

Pore Space

The volume of soil mass that is not occupied by soil particles is known as pore space. The pore space is usually occupied by air and water. In the pore spaces, the plant roots grow and exit. It directly controls the amounts of water and air in the soil and thus indirectly controls plant growth and crop production. In general two sizes of pores are recognized *i.e.,* Macro or non- capillary and Micro or capillary.

Soil Colour and Composition

(*a*) *Black or dark grey:* The variations from black to dark grey colour of soil are mainly due to organic matter.

(*b*) *Brown colour:* This is the most common soil colour and is due to a mixture of the organic matter and iron oxides.

(*c*) *Red – yellow colour:* Red colour is associated with anhydrate ferric oxides, whereas yellow colour indicates some degree of hydration.

(*d*) *White colour:* Silica and lime generally impart white colour.

(*e*) *Bluish and Greenish colour:* Due to the presence of ferrous compounds in ill-drained soil.

Importance of Soil temperature

Effect of soil temperature on soil microorganism: Soil temperature controls the rate of microbial process involved in mineralization of nitrogen, composition of organic matter, nitrogen-fixation etc. There is lowest microbial activity where soil temperature is below 5°C and above 54°C.Optimum temperature is between (25 – 35)°C.

Effect of soil temperature on the decomposition of soil – organic matter: The higher the temperature the more rapid is the decomposition.

Effect of soil temperature on absorption of soil water: Very low and very high temperature affects the absorption of soil water by the roots of plants.

Effect of soil temperature on nutrients availability: The solubility of food nutrients and their availability are considerably influenced by soil temperature.

Effect of soil temperature on seed germination: If the temperature is too low, the seed fails to germinate or germinate very slowly. If it is too high, seed may be injured.

Effect of soil temperature on plant growth: During plant growth physiological processes are influenced by temperature.

Effect of soil temperature on soil formation: Process of weathering and soil development are accelerated where there is an increase in soil temperature.

Effect of soil temperature on plant diseases: Development of diseases is also related with soil temperature.

Soil Depth

Tea grows best in deep soil. A soil can be called deep if it is 90cm or more in depth, it is moderately deep when the depth is 50-90cm and only shallow when the depth is 25-50cm.

Different management practices are used for varying depths of soils under mature tea, young tea and for replanting. If very shallow (0-25cm), or merely shallow (25-50cm) depth occurs in soils under mature tea area then the following steps are to be taken:

(a) Maintenance of good shade

(b) Split application of fertilizer

(c) Irrigation at frequent intervals with lower amount

(d) Shorter pruning cycle

(e) Chemical weed control

Soil Moisture

Soil Moisture has effect on physical and chemical reaction of the soil. The maximum availability of water to the plant is when the soil holds water at around 0.33 bar. However, there is very little water available for plants at 15 bar suction. The soil moisture between these two limits is the so called "Available Water". Optimum moisture is required for proper plant growth.

Soil Air

Adequate presence of air is necessary for proper growth of plant. Inadequate soil aeration adversely effect root growth, root respiration and other essential biological activity in the soil. Depending on soil structure soil have pore space of 30-60 per cent.

In the total pore space about 1/3rd is occupied by air and 2/3rd is occupied by water is favorable for plant growth.

Like atmospheric air soil air is mainly composed of nitrogen and oxygen. The percentage composition of forest soil air is approximately as follows-

$CO_2 = 0.25\%$, $O_2 = 20.6\%$, $N_2 = 79.2\ \%$

Proper drainage and tillage work help in proper aeration of the soil.

Besides proper plant and root growth soil air helps in increasing microbial population and its activity, easy absorption of water and nutrients by the root.

Tilth

Tilth refers to the "Physical condition of the soil in relation to plant growth" and hence must include all soil physical condition eg. moisture content, degree of aeration, rate of water infiltration, drainage and capillary water capacity. These all physical conditions are responsible for influencing crop development.

Soil Profile

It is a vertical section of the soil through all its horizons and it extends up to the parent material. With the increase in depth changes in texture and structure of the soil is noticed. The soil column in most cases is composed of a series of layers which are parallel with the soil surface and are called soil horizons.

Oi – Organic Slightly decomposed

Oe – Organic Moderately decomposed

Oa – Organic Highly decomposed

A – Mineral. Mixed with humus, dark coloured

E – Horizon of maximum accumulation of silicate clays, Fe and Al oxides

AB or EB – Transition to B, More likely A or E than B

BA or BE – Transition to A or E, more likely B than A or E

B – Most clearly expressed portion of B horizon

BC – Transition to C, more likely B than C

C – Zone of least withering, accumulation of Ca, Mg carbonates , cementation, sometimes high bulk density

R – Bed Rock

Usda Taxonomy of Soil Surface Horizon

Mollic (A): Thick, dark coloured, high base saturation, strong structure.

Umbric (A): Same as Mollic except low base saturation.

Ochric (A): Light coloured, low organic content

Histic(O): Very high in organic content

Anthropic (A): Man modified Mollic like, high in available Phosphorus

Plaggen (A): Created by year of manuring.

Sub-surface Horizons

Argillic (Bt) : Silicate clay accumulation.

Natric (Btn) : Argillic , high in Sodium, columnar or prismatic structure.

Spodic (Bhs) : Organic matter, Fe and Al Oxide accumulated.

Cambic (B) : Changed or altered by physical movement or by chemical reactions.

Agric (A or B) : Organic or altered by physical movement or by chemical reactions.

Oxic (BO) : Highly weathered, mixture of Fe, Al oxides.

Duripan (M) : Hard pan, strongly cemented by silica.

Fragipan (X) : Brittle pan, loamy textured, weekly cemented.

Albic (E) : Light coloured , clay and Fe, Al oxides mostly removed.

Calcic (K) : Accumulation of $CaCO_3$ or $CaCO_3 \cdot MgCO_3$.

Gypsic (Y) : Accumulation of gypsum.

Salic (Z) : Accumulation of Salts.

Classification of Soil Micro Organism

Bacteria

Heterotrophic Bacteria: These bacteria depend upon organic matter or living body for their food. These bacteria decompose the organic wastes and dead bodies of plants and animals and release chemical elements locked up in them. Functionally, they bring about mineralization of organic matter through hydrolysis and oxidation and release nitrogen, phosphorus and other nutrients in available forms to plants. They also fix atmospheric nitrogen in the soil.

A) Nitrogen – fixing bacteria

(i) Symbiotic, e.g., *Rhizobium*

(ii) Non –symbiotic

a) Aerobic, e.g., *Azoctobactor*

b) Anaerobic,e.g., *Clostridium*

B) Non- nitrogen fixing bacteria

(i) Aerobic,e.g., ammonifiers

(ii) Anaerobic,e.g.,denitrifiers

Autotrophic Bacteria: These bacteria manufacture their food (complex compounds) by using very simple inorganic substances. These bacteria oxidize ammonia, nitrate sulphur, manganese, iron, carbon monoxide, hydrogen and methane, so that plant can utilize these substances.

(i) Nitrifying bacteria

(ii) Iron bacteria

(iii) Hydrogen bacteria

(iv) Sulphur bacteria

(v) Manganese bacteria

(vi) Methane bacteria

(vii) Carbon monoxide bacteria

Fungi

(i) Yeast

(ii) Mould

(iii) Mushroom fungi

Algae

(i) Green.

(ii) Blue – green.

(iii) Diatoms.

Animals

Micro animal

(a) Protozoa

- (i) Paramecium
- (ii) Euglena
- (iii) Amoeba

(b) Nematodes

Macro animal

(i) Earthworms

(ii) Mites

(iii) Centipedes

(iv) Millipedes

(v) Snails

(vi) Mice

(vii) Termite

Function of Soil Microorganism

Beneficial role of Bacteria

Change and decomposition in organic matter: The organic materials when incorporated in the soil are attacked by the micro-organisms. They break up the various constituents of the organic matter of the organic materials and convert them into new substances. These are of the following types:

(i) *Ammonification*: The change of organic nitrogenous compounds into ammonia is called Ammonification. It is occurs as a results of action of enzymes produced by micro-organism in the presence of air.

Protein ——— Polypeptides——— Amino acid——— Ammonia

(ii) *Nitrification*: The conversion of ammonia to Nitrate(NO_3) is known as nitrification. Autrophic bacteria performed this transformation. It is an aerobic process.

Ammonia—— Nitrosomonas bacteria ———Amino acids ———Nitrite— —Nitrobacter——— Nitrate

Decomposition of simple Products: The more common simple products resulting from the activity of the soil micro-organisms are as follows:

Carbon: (CO_2, CO_3^-, HCO_3^-, elemental carbon).

Sulphur: S, H_2S, SO_3^{--}, SO_4^{--}, CS_2.

Phosphorus: $H_2PO_4^-$, HPO_4^{--}.

Others: H_2O, O_2, H_2, H^+, OH^-, K^+, Ca^{++}, Mg^{++} etc.

Mineralization of Organic Sulphur: Autotrophic bacteria oxidized sulphur into sulphate form.

Sulphur (S) + Oxygen ($2O_2$) — Enzymic oxidation——— Sulphate (SO_4).

Mineralization of Organic Phosphorus: Organic phosphorus compounds are mineralized by the action of microorganism into inorganic phosphorus.

Organic Phosphorus ———Microorganism-——— $H_2PO_4^-$ (available phosphorus) or HPO^{4-}*Fixation of Atmospheric Nitrogen:* The conversion of elemental nitrogen to readily available form by nitrogen – fixing microorganism is called biological nitrogen fixation. The nitrogen fixed by microorganisms in the soil is known as Bio – fertilizers. The nitrogen fixing microorganisms are: (i) Bacteria (ii) Algae and (iii) Mycorrhizae.

Algae: Blue- green algae are capable of fixing atmospheric nitrogen to the soil. The main genus of algae which fixes nitrogen are *Anabaena, Nostoc,* and *Cylindrospermum.*

Mycorrhizae: Mutually beneficial association between certain fungi and roots of higher plants is called Mycorrhizae (fungus roots). Mycorrhizae are divided into two types: (a) ectomycorrhizae(mainly *Amanita*, *Boletus* etc.) and (b) endomycorrhizae (this group also called *vesicular – arbuscular mycorrhizae* or VAM).

Soil Consistency

It is represents at varying moisture conditions, the degree and kind of cohesion and adhesion of soil material. Cohesion refers to the attraction of substances of like characteristics such as that of one water molecule for another. Adhesion is the attraction of unlike materials, for example, attraction between soil and water molecule.

Consistence of soil depends on the texture, colloids, structure and especially the moisture content of soils. The consistence of soil is generally described at three soil moisture levels:

- **Consistence when soil is wet (Moisture content at field capacity):** is described in terms of stickiness and plasticity.

Stickiness: The quality of adhesion to other objects is called stickiness. The degree of stickiness is identified by the following terms: non- sticky, slightly sticky, sticky and very sticky.

Plasticity: is the capacity to be moulded. When stress is applied, the shape changed and after removing the stress it keeps the changed shape. Terms used to describe plasticity are: non- plastic, slightly plastic, plastic and very plastic.

- **Consistence when soil is moist (Moisture content between air dry and field capacity):** it is described in the following terms:

  Loose: non-coherent.

  Very friable: coherent but very easily crushed.

  Friable: easily crushed.

  Firm: Crushed under moderate pressure.

  Very Firm: Crushed only under strong pressure.

  Extremely firm: resists crushing between thumb and forefinger.

- **Consistence when soil is dry (air dry):** For dry condition of soil consistence is characteristic by hardness.

  Loose: non coherent

  Soft: breaks under slight pressure between thumb and forefinger to a powdery mass.

Slightly hard: breaks under moderate pressure.

Hard: breaks with difficulty under pressure.

Very hard: can't be broken in thumb and forefinger.

Extremely hard: extreme resistance to pressure; can't be broken in the hand

Role of Physical Properties in Soil Fertility

Importance of Texture: Sandy soils have poor fertility and very little water holding capacity so only few crops can be grown such as potato, groundnuts etc. Clay soils have high fertility and very high water holding capacity. Rice, jute, sugarcane can be grown in this soil. Silt (loam) soils have medium fertility and water holding capacity. They are suited to every kind of crop such as wheat, maize, pulses, vegetables, sugarcane etc.

Importance of Structure: Platy structure normally results in water logging. This structure is not good for cultivation .Whereas sphere – like structure helps in drainage and provides optimum aeration, water holding capacity etc. This structure is considered best for cultivation.

Importance of Density: Soils with low bulk density have favourable physical conditions.

Importance of Porosity: For ideal conditions of aeration, permeability and water holding capacity, a soil should have an equal amount of macro and micro pores.

Importance of Consistence: Optimum soil consistence increases the water holding capacity and plant food material. Thus, increasing the fertility of soil.

Importance of Soil Colour: Soil colour are indirectly related with soil fertility:

(a) Relation of soil colour with temperature: Directly the soil colour influences soil temperature to some extent. The dark colored soils absorb more heat than light colored soil.

(b) Relation of soil colour with organic matter: organic matter imparts black to grey tinges. More quantity of organic matter content improves the soil structure.

(c) Relation of soil colour with drainage: Bluish and greenish colour of soil indicate water logging. Need of drainage can be determined by the colour variations in the soil.

(d) Relation of soil colour with parent material: soil colour may indicate their origin from parent material.

Note: The information on Soil Science have been derived from the books (i) Fundamentals of Soil by V.N.Sahai (ii) A Text Book of Soil Analysis-T.C.Baruah & H.P. Borthakur).

Influence of Soil and Other Parameters on Soil Applied Fertilizer use Efficiency

1. **Soil Texture**: Lighter the soil texture i.e. higher the sand content greater is the porosity but lesser is the retention of nutrients (N, K, Ca, Mg, and Al) and water.

Suggestions

(a) Enrich organic matter as much as possible by extraneous application.

(b) Stress on more foliar application as long as the soil is enriched with desired organic matter content.

(c) Avoid application of more organic matter in soil having high silt content. This will reduce the porosity of soil.

2. **Soil Structure** : Generally poor in low land under cultivation of paddy and planted in non-rehabilitated area

Suggestions

(a) Light forking in Light Pruned years

(b) Apply dolomite as a source of calcium

(c) Get rid of water logging by good drainage

(d) Enrich organic matter content

(e) Check the root distribution as a sample (uproot and check). If it is poor, give more stress on foliar application till the root develops properly.

3. **Organic Matter Content** – Higher the organic matter content better is the fertilizer use efficiency. Target should be to get a organic carbon more than 2 per cent. Lesser the organic matter content, higher the loss of fertilizer through leaching, lesser the sustained available nutrient content, higher the requirement of fertilizer to compensate the loss.

Suggestions

(a) Add any source of organic matter (cattle manure, oil cakes, decomposed tea waste, decomposed shrubs, weeds, vermicasting etc.)

(b) Grow Guatemala grass / Citronella in vacant patches until it is in filled.

4. Soil pH

pH: Below 4.5 – Not suitable for tea cultivation

4.5 to 5.5 – Suitable

Above 5.5 – Not suitable

Suggestions

(a) If pH is high, adjust with aluminium sulphate or pyrites

(b) If pH is low, adjust with dolomite

5. **Fertility Status** : Soil fertility depends on nutrient content of the soil and soil micro organisms.

Suggestions

(a) Every tea estate should have soil report of every section.

(b) Soil testing should be done during every LP year

(c) Addition of organic manure along with extraneous application of microbes like *Azotobacter, Azospirillum, Bacillus megatherium* etc in the pruned years will help improvement of soil fertility.

6. Drainage Status: Improved drainage system can assure higher fertilizer use efficiency.

Suggestions

(a) Improved drainage condition is the top priority

(b) Till drainage is improved, foliar application of urea and potassium plays a great role.

(c) If soil phosphate status is low and phosphorus is bound due to drainage problem, spray soluble phosphate (DAP) instead of applying ground phosphate.

Chemical Properties

1. Soil pH
2. Ion Exchange
3. Available Nitrogen
4. Available Phosphorus
5. Available Potassium
6. Available Calcium
7. Available Sulphur
8. Available Magnesium
9. Available Zinc
10. Available Boron

Soil pH

Soil pH is defined as the negative logarithm of hydrogen ion concentration. It indicates the acidity or alkalinity of soil. pH of water is 7 which is neutral. pH range of soil may usually vary from 3.5 to 10.5. Soil pH above 7 indicates soil is acidic and above 7 indicates soil is alkaline. Tea grows well in pH between 4.5 to 5.5.

A change in of one unit of pH indicates 10 times increase or decrease in hydrogen ion concentration.

Reasons of low or high pH are:

- Too much use of urea, even in foliar spray it is used.
- Garden near to dolomite mine.
- Inadequate drainage.
- If pH is low (below 4.5) it can be increased by applying dolomite @ 2 tons/ha in Nov-Dec in coarse textured soil and @ 3 tons/ha in fine textured soil.
- If pH is high (between 5.6-5.8) in can be lower by applying Aluminum Sulphate or Sulphur Pyrites.
- Aluminum Sulphate is more costly but where organic matter is less there Aluminum Sulphate is used instead of being costly because Pyrites need microbial activity to release.

Ion Exchange

Ions may be positively charged and negatively charged. Ion exchange is a reversible process by which cations and anions are exchanged between solid and liquid phases. The exchange of cations and anions are known as cation exchange and anion exchange respectively.

Role of Various Nutrients Available Nitrogen

Nitrogen in soil existed mostly in the form of compounds resulting from decaying vegetable matter and was present in complex and simple organic compounds, as well as in the form of ammonia, nitrites and nitrates.

The major source of nitrogen in soil is organic matter.

Generally the nitrogen content progressively depleted over the period of cropping. In Jorhat area, the depletion between virgin jungle soil and old tea soil was more pronounced after forty years of cropping.

Application of the chemical significantly influences the soil nitrogen, carbon, and C:N ratio. A higher dose of nitrogen (135kg/ha) resulted in better nitrogen, carbon and C:N status of the soil than a lower dose of nitrogen (45 kg/ha).

Total nitrogen of the top soil was found highly crop related with yield. Pruning litter and Nitrogen fertilizer application of up to 200 kg/ha helped the residual build up of soil Nitrogen and increased shoot Nitrogen.

Nitrogen is not easily available to the plant directly. There are some nitrogen fixing bacteria like *Nitrobactor, Azotobacter, Nitrosomonas* that fix the atmospheric nitrogen through Nitrogen Fixation.

Available Phosphorus

The optimum range of available Phosphorus should be 10 to 20 ppm.

Applied as (i) DAP- Foliar Spray

(ii) RP- Acid Soluble

(iii) SSP- Water Soluble

DAP should be given as 1% foliar spray (Total 22 kg/ ha/ year.)

After huge attack of looper, red spider mite, hail storm should go for spray of DAP.

DAP and Zn are incompatible so DAP should not be applied with Zn as a cocktail..

Blanket Dose: 50 kg P_2O_5 /ha/year.

Phosphorus

1. Immobile in soil as well as in leaf.
2. It will not leached out.
3. Rock phosphate is slow releasing, it is persistent in soil releases gradually.
4. RP is not water soluble so during heavy rain also it will not lost.
5. If pH is more than 5.5 then SSP should be applied and if pH is low then RP should be applied.
6. Phosphate has a tendency to bind with Al, Ca, Fe. If once bind then it is not available to the plant.

7. *Bacillus megatherium* and VAM activates phosphorus uptake of the plant.
8. Phosphate is required 30 g/plant.
9. Placement method is better in case of phosphorus.
10. In Feb.-March along with organic manure 30 g of RP is given per pit and it is placed between 2 plants.
11. It helps in rooting of the plant.

Potassium

1. Potash controls the stomatal opening and give strength to the plant.
2. Due to potash deficiency plant become weak, short, weepy, thin, top surface becomes breakable, full of die back will be there.
3. Potash is given as Muriate of Potash. MOP is a luxury nutrient because it has no toxicity
4. Doesn't require microbial action for activation
5. In higher dose also it will not affect the plant

As per latest finding, when lower pruning is done potash requirement increases.

If less than 45cm, $N:K_2O$ :: 1:2 (In case of RP,CP)

If 45 cm to 60 cm, $N:K_2O$ :: 2:3

If above 60 cm, $N:K_2O$ :: 1:1

During drought condition when stomatal conductance increases then foliar application of potash should be given starting from December to April (fortnight).

During rainy season when the area is submersed then foliar application of potash should be given.

Sulphur

Sulphur is an essential plant nutrient having three amino acids (cystine, cystiene and methionine) which serve as the precursor of VFC synthesis.

Optimum range of Sulphur in soil is more than 30ppm.

It is applied as elemental sulphur, @ 50 kg in 2 splits ; Pre & Post monsoon.

Sulphur should not be applied during rainy season. Because it reacts with water and form H_2S which has a toxic effect to the plant.

Zinc

Zinc is an essential plant nutrient absorbed by roots as Zn^{++} .

About 62% of soils studied had Zn between 3ppm-15ppm and another 29% between 15ppm-27ppm. Only 4% of the soils had Zn above or below the limits of 3 ppm to 27 ppm.

Available Zinc of soils did not vary much vertically up to a depth of 120cm.

About 4.6per cent of the total zinc remained in available form in tea soils, and both total and available zinc were linearly correlated (r2 = 0.66). The total Zn content of tea soils vary from 170 ppm to 570 ppm.

Boron

It is one of the essential micronutrients absorbed by roots as $H_2 BO_3^-$ or $H_3 BO_3^-$ ·Total Boron content varied from 8.80ppm to 25 ppm, and available Boron from 0.88 ppm to 0.44 ppm. Available Boron content of tea soils is low being 1 per cent of the total in the South bank, and between 1.5% to 1.8% in the North bank.

Available Boron has a strong positive correlation with total Boron quantity.

Soil availability of Boron decreases significantly with increasing level of Ca application.

Role of Nitrogen

1. Nitrogen is a part of all living cells and is a necessary part of all proteins, enzymes and metabolic process involved in the synthesis and transfer of energy. Affect energy reactions in the plants.
2. Essential components chlorophyll, which is responsible for photosynthesis. Imparts dark green colour to tea plants.
3. Present in nucleotides, phosphatides, alkaloids, enzymes, hormones, vitamins etc.
4. Necessary for formation of amino acids, the building blocks of protein.
5. Promotes leaf, stem and other vegetative growth.
6. Aids in production and use of carbohydrates.

Pathways of Nitrogen

Urea (46%N) $[CO(NH_2)_2]$

> Hydrolyses by urease enzymes

> Ammonium carbonate $(NH_2)CO_3$

> Dissociation / Mineralization > $NH_4 + H_2CO_3$

> $CO_2 + H_2O$ – Adsorbed by clay mineral and organic matter

> NO_2 (unstable)

>NO_3 (mobile)

> Taken up by root or leached out

>Amine (NH_2) (Hydrogen Reductase enzyme)

> Amino acids

>Protein (cell development)

Bad Effect of Excessive or Low Nitrogen

1. Less application will slowly exhaust soil organic bound
2. Less application will tell upon yield potential in fertile soil but ultimately hamper crop.

3. Excess application will increase $NO_3 - N$ fraction in the plant, thus will weaken cell wall and invite more pests.
4. Excess application will induce NH_4 + - N toxicity in water logged soil, thus will affect the feeder root.
5. Excess application will induce Calcium deficiency and reduce root growth.

Role of Phosphorus

1. Phosphorus is an essential part of the process of photosynthesis.
2. Involved in the formation of all oils, sugars, starches etc.
3. Constituents of nucleic acids, phytin, and phospholipids.
4. Encourages blooming and root growth.
5. Promotes early root formation and growth.
6. Responsible for cell division and enlargement.
7. Helps with the transformation of solar energy into chemical energy; plants survive harsh winter conditions.
8. Increase water-use efficiency and hastens maturity.

Role of Potassium

1. Vital to photosynthesis, essential to protein synthesis.
2. Increases water use efficiency and turgidity ; regulates stomatal function
3. Increases Nitrogen metabolism, and breakdown of carbohydrate & translocation of carbohydrate from leaf to storage organs; translocation of starch.
4. Activates enzymes and controls their reaction rate.
5. Counteracts the injurious effect of excess Nitrogen.
6. Imparts increased vigour and disease resistance to plants.
7. Improves the quality of seeds and fruit.
8. Improves winter hardiness.

Pathways of Potasium

Dissolves /Dissociates rapidly in soil

>Unavailable (fixed in clay minerals).

>Adsorbed in clay.

>Soil solution.

>Taken up by plant / leached out under high rainfall and low pH.

Bad Effect of Excessive or Low Potassium

1. Less application in heavy soil will reduce Nitrogen metabolism in the plants and induce pest and disease attack, make the plant susceptible to drought.

2. Excess application will induce loss of Potash through leaching
3. Potassium is known as LUXURY NUTRIENT hence excess application is not very problematic

Role of Magnesium

1. Key element of chlorophyll, proteins, enzymes, and certain volatile compounds.
2. Very much essential for photosynthesis.
3. Helps in synthesis of carbohydrates, protein, fat and vitamins.
4. Regulates uptake of other nutrients.
5. Increase iron utilization in plants.
6. Activator and component of many plant enzymes.
7. Improves utilization and mobility of phosphorus.

Role of Calcium

1. Constituent of cell wall structure, provide strength to the plant.
2. Promotes early root development and growth.
3. Neutralises poisonous organic acids and alkali salts.
4. Activates root tips.
5. Improves uptake of other plant nutrients.
6. Involved in nitrogen metabolism.
7. Reduces plant respiration.
8. Stimulates microbial activity.

Role of Sulphur

1. Constituents of Sulphur bearing amino acids like cystine, cystiene and methionine.
2. Acts as a precursor of Volatile Flavour Constituents.
3. Promotes nodule formation.
4. Essential plant food for production of protein.
5. Promotes activity and development of enzymes and vitamins.
6. Helps in chlorophyll formation.
7. Helps with vigorous plant growth and resistance to cold.

Role of Zinc

1. Constituents of several enzyme system, which include alcohol dehydrogenase, lactic acid dehydrogenase, glutamic acid dehydrgenase, carboxypeptidase.

2. Influences the formation of growth hormones in plants.
3. Zinc is required in the synthesis of tryptophane, a precursor of Indole Acetic Acid. Formation of this auxin is influenced by Zinc.
4. Zinc has a possible role in plant metabolism involved in starch formation.
5. Associated with water uptake and water relations in plants.
6. Regulate composition of sugar.
7. Necessary for chlorophyll production.
8. Necessary for starch formation.

Role of Iron

1. Promotes formation of chlorophyll.
2. Acts as a oxygen carrier.
3. Reactions involving cell division and growth.

Role of Boron

1. Keeps Calcium in soluble form and increases its mobility.
2. Helps in absorption of Nitrogen and Carbohydrate.
3. Constituent of cell membrane and essential for cell wall formation.
4. Precipitates excess cations.
5. Helps in translocation of sugars.
6. Essential for germination of pollen grains and growth of pollen tubes.

Role of Manganese

1. Supports movement of iron in the plant.
2. Helps in chlorophyll formation.
3. Acts as catalyst in oxidation reduction reactions within the plant tissues.
4. Helps in respiration and protein synthesis.
5. Increase the availability of Phosphorus and Calcium.

Role of Copper

1. It forms many compounds with amino acids and proteins in the plant.
2. Acts as electron carrier in enzymes.
3. Catalyzes several plant process.
4. Major function in photosynthesis.
5. Increase in sugar content.
6. Aids in root metabolism and helps in the utilization of proteins.

Role of Molybdenum

1. Required to form the enzyme "nitrate reductase" which reduces nitrates to ammonium in plant.

2. Aids in the formation of legume nodules.
3. Needed to convert inorganic phosphates to organic forms in the plant.
4. Helps in the use of nitrogen.

Deficiency Symptoms

Nitrogen

1. The colour of the flush and young leaves become light green
2. Older leaves turn brown and fall.
3. The leaves under bright sunshine looses colour very rapidly.

Phosphorus

1. The leaves become dull and matty.
2. Shine of the leaf is lost.
3. A bluish tint develops at the later stages.
4. Die back of stems possible.
5. During phosphate deficiency, stems die back from the ends which have been cut for pruning.

Potassium

1. The tip and the apical margins becomes black coloured.
2. Sometimes leaves turn red bronze in colour.
3. On shaking a Potassium deficient bush, many leaves fall off.
4. Maintenance canopy becomes scanty.
5. Brown blight *Colletotrichum camelliae* and grey blight *Pestalozzia theae* are associated with Potash deficiency.
6. The bark becomes silvery white.

Magnesium

1. Deficient leaves develop a characteristic marginal yellowing which leaves a dark green V – shaped area.
2. Old leaves are affected first.
3. Inverted V shaped dark green patches develops on the leaf surface.

Calcium

1. The stems shorten, the leaves do not grow to normal size.
2. The leaves turn yellow and curl backwards.
3. Leaf tips and edges turn black , the leaf distorts and cracks, then fall off.

Sulphur

1. Deficient leaves become bright yellow between the veins, which remain green.
2. New leaves become smaller and the inter nodal distance is reduced.

3. The leaves scorch and fall off and stems die back from the tips. The condition is known as tea yellows.

Zinc

1. Resetting of leaves near nodal region.
2. Sickle shaped leaves.
3. Wavy edge leaf margins often go yellow.
4. Short internodes.

Boron

1. This deficiency is also very rare.
2. The apical bud fails to grow and cork develops at the upper side of the leaf petiole.

Manganese

1. The deficiency is very rare, occurring occasionally on soil of very high pH.
2. Leaves turn pale yellow on the edges and develop a mottling of red brown sports.

Copper

1. Slightly darker leaves.
2. The main effect of Cu – deficiency is to inhibit fermentation in the factory.
3. Fermentation takes longer than usual and the leaf becomes dark brown rather than bright orange.

Photographs of Deficiency Symptoms

Zinc, Magnesium, Nitrogen and Potash deficiency

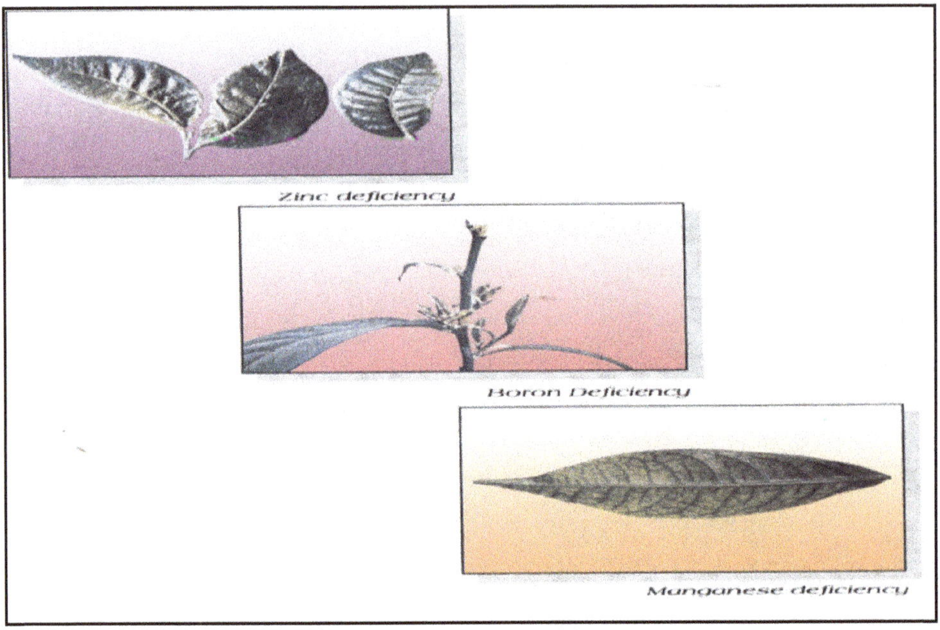

(Courtesy : Tocklai Experimental Station & T.Stanes & Co.)

Acute Magnesium deficiency

Nutrient Content Of Fertilizers

1. NITROGEN : Urea (46 % N); Ammonium Sulphate (16 % N)
2. POTASSIUM : MOP (60 % K_2O)
3. PHOSPHATE : SSP (16 % P_2O_5, 19.6% Ca, 11-12 % S) ; Rock phosphate (20-22 % P_2O_5, 33 % Ca) ; DAP (46 % P_2O_5, 18 % N)
4. SULPHUR : Gypsum (18.6 % S) ; Wettable Sulphur (80% S)
5. ZINC : Zinc Sulphate($ZnSO_4$, $7H_2O$) (22 % Zinc, 11 % S) ; Zinc EDTA (10 % Zn)
6. MAGNESIUM : Magnesium Sulphate ($MgSO_4 . 7H_2O$)(9.4 % Mg, 13 % S)
7. BORON : Borax (Na-tetraborate) (10 % B)(soluble in cold water ; Boric acid (H_3BO_3) (10 % B) (insoluble in cold water)
8. DOLOMITE : 21 % Ca, 12 % Mg
9. PYRITE : Pyrite (FeS_2) (52 % S, 46 % Fe)
10. ALUMINIUM SULPHATE : 22% Al, 26 % S

Method of Soil Sampling

The importance of having a truly representative sample can be very well realized from the fact that only a few grams of soil of the huge mass of the soil in the field are analysed in the laboratory and the analytical data are expressed per hectare basis and the data are applicable to the whole area. It is therefore important that samples are representative of the soil of the area under investigation; unless this is ensured, sampling may be the greatest source of error in the whole soil testing programme. No amount of improvement in the laboratory techniques can compensate for a poorly collected sample.

The following points are to be taken into consideration for collecting representative soil sample.

1. Soil sample should not be collected from a field within 90 days after the application of organic manures or water soluble phosphatic fertilizers and 30 days after the application of nitrogenous and potassium fertilizers. Avoid soil sampling from the spots where rock phosphate has been applied by placement at least for one year, as it solubilises very slowly depending on climatic conditions and availability of moisture.
2. Do not sample unusual areas. Avoid areas near road margins, labour line sites, cattle shed neighbourhood, marshy spots, near shade trees, compost piles, bunds, channels and other non-representative locations.
3. At the time of sampling, variation in slope, soil colour, texture, crop growth and drainage should be taken into account and separate sample should be collected from each area.
4. The field should be divided into units across the slope and never along the slope. The size of a sampling unit may be taken as 0.5 to 2.0 hectares depending on the variations.

5. From each sampling unit, one set of composite soil sample from surface soil (0 – 22.5 cm) and subsoil (22.5 – 45.0 cm) may be mixed from three to five individual soil samples collected in a zigzag way and not in a straight or circular line.
6. Before taking the soil sample, scrape away the surface litter but the surface soil should not be removed.
7. The soil sample should be collected between tea rows.
8. Soil samples can be drawn with any of these hand implements namely (i) soil tube auger (ii) screw type auger (iii) post hole auger (iv) spade) (v) crow bar (vi) gardening tool.

For sampling of soft moist soil, the tube auger, spade. Gardening tool is quite satisfactory. A screw type auger may prove convenient on hard dry soil, while the post hole auger is useful for sampling excessively wet or soils from marshy area. If spade or gardening tool is used a "V- shaped" cut may be made up to 45.0 cm depth and a uniformly (1.5 cm approximately) thick slice is taken out from top to bottom firstly of surface soil (0 – 22.5 cm) and then sub soil (22.5 – 45.0 cm) from both sides of "V – shaped" cut repeatedly four to five times. The soil samples from both depths should be collected separately and while collecting soil sample from sub soil, care should be taken that surface soil does not mix with the sub soil.

9. The individual soil samples collected in this manner from different places in a sampling unit should be thoroughly mixed by hand on polythene sheet or metal tray; the bulk quantity of soil sample is reduced by quartering to make one composite sample. Quartering may be done by mixing the sample well and dividing it into four equal parts, then rejecting two opposite quarters while mixing the remaining two portions. The quantity of the sample may be reduced to about 250 g by repeating this process. The composite sample thus obtained is sent for analysis to the laboratory.
10. If the sail sample is wet at the time of sampling, it should be dried in shade before packing, but should not be heated.
11. The soil samples for nitrite (NO_2) analysis do not require drying in shade as it should be analysed in fresh soil samples. After collection, the soil samples may be stored in the refrigerator, if there is time interval between soil sampling and forwarding to the laboratory for analysis.
12. Bags used earlier for packing fertilizer, insecticide or fungicide should not be used for pacing soil samples. It is suggested to use new plastic bags for packing soil samples. The ideal thickness of polythene film is about 0.13 mm of the polythene bag for packing soil sample.
13. The sample should be packed with labels inside and outside. The labels should be written with black ball pen. Do not use ink or copying pencil for writing labels.

14. Each soil sample should contain a label bearing the following information.
    (a) Name of the estate (b) Name of the division and filed number (c) Depth of sampling (d) Date of sampling (e) Purpose of sampling
15. The soil samples should be sent to the laboratory for analysis with a packing note having details of soil samples, the analysis required and the nature of problem for which advice is sought.

Soil sample for Tea Board Subsidy

Top (0 – 22.5 cm) and bottom (22.5 – 45.0 cm) soil should be collected at the rate of one set of sample for every 2 ha of proposed area.

(Ranganathan,V&Natesan, S. (1987). Manuring of tea- Revised Recommendations. Handbook of Tea Culture, Section 11. UPASI Tea Research Institute).

Fertilizer Programme

A bit of planning is required for making a fertilizer programme in a tea garden. Before doing it, following inputs are required from each section of a tea garden:

(i) The Productivity (Yield per hectare) - Made Tea Kilograms per hectare per year (MTKHY)

A mean of preceding five years excluding the pruned year should be taken into consideration . This data is required because when 100 kg of green leaf is harvested, about 3.5 kg of Nitrogen is removed from the soil, hence it has to be replenished by exogenous application of Nitrogen. The dose of Nitrogen is ascertained on the basis of yield potential of a section.

(ii) Soil analysis report – The soil testing report of available Nitrogen, Organic Carbon, available Potassium, available Phosphorus, available Sulphur, pH is required. The dose of Potassium should be fixed on the basis of Potash status in a tea section. pH is the negative logarithm of Hydrogen ion concentration. In the pH scale of 0 to 14; below 7 it is acidic and above 7 it is alkaline. At 7 pH , the reaction is neutral because at 7 pH , the concentration of Hydroxil ions (OH ions) is equal to Hydrogen (H) ions. If the pH is low (*i.e.* acidic) , dolomite @ 2 tonnes per hectare is applied and if the pH is high (*i.e.* alkaline) Iron/Sulphur pyrites or Aluminium Sulphate is applied.

(iii) Pruning /Skiffing programme of the last few years. In the pruned sections, requirement of Potassium increases.

(iv) Shade status – In case of poor shade status, the possibility of obtaining nutrients through loppings and leaf fall will be less, hence more Nitrogen should be suggested in those kind of sections.

Calculation

As per the recommendations of TRA, the ready reckoner for Nitrogen is as follows:

YTD Doses Recommended by TRA

Year of Planting	N (kg/ha)	YTD (kg/ha)	Method of Application
0 Year	20-40	200-400	Ring in 2-3 splits
+1 Year	80-100	800-1000	Ring in 4 splits
+2 Year	100-120	1000-1200	Ring in 4 splits
+3 Year	120-140	1200-1400	Ring in 4 splits
+4 Year	140-150	1400-1500	Strip in 2 splits

Mature Tea Manuring (Plains) Doses Recommended by TRA

Yield range (MTKHY) Pruning Cycle average	N(kg/ha)	P_2O_5 (kg/ha)	K_2O (kg/ha) Soil available Potash status		
			Low (Less than 60 ppm)	Medium (60-100 ppm)	High (More than 100 ppm)
Upto 1500	Upto 90	20	Upto 90	Upto 70	Upto 50
1500-2000	90-110	20-30	90-110	70-80	50-70
2000-2500	110-140	30-50	110-140	80-120	70-100
2500-3000	140-150	50	140-165	120-140	100-120

Another method of calculation for manuring in mature tea sections is as follows :

Upto 2000 kg yield per hectare per year, the dose of Nitrogen should be 160 kg. For every 100 kg increase, 4 kg should be added. For example , if in a section the yield is 2100 kg /hectare, the dose will be 160 + 4 = 164 kg /hectare.

Deduction for Vacancies and Drains/ Road etc.

The vacancy percentage (in case tea bushes have died due to some reason) should be calculated manually and the exact percentage should be deducted from the calculated requirement. Suppose in a section the vacancy percentage is 5 per cent and the total requirement of Nitrogen in that year is 6500 kg., the quantity will stand as 95per cent of 6500 = 6175 kg. A school of thought believes that there should be no deduction for vacancies because the fertilizer will go to the soil of the vacant area, which will be of some use to the tea plants in future. Generally , for drains and roads 4 per cent is deducted.

Example

What should be the fertilizer doses in a section having following statistics?

Location : Dooars (West Bengal)

Area : 20.5 hectares

Style of Planting : Staggered Single hedge

Spacing : 105 cm X 60 cm

Vacancy : 4 % (due to death of tea plants)

Organic Carbon : 1.5 %

Potash status : Optimum

Sulphur status : Low

pH – 4.5

Productivity (MTKHY) : 3500 kg (Mean of one pruning cycle)

Pruning / Skiffing status in the year of application : Light Skiff (90cm from the ground level)

Nitrogen requirement

For 2000 kg – 160 kg

For 1500 kg – 60 kg

Total - 220 kg per hectare

In 100 kg Urea – 46 kg of Nitrogen is available

So, for 220 kg Nitrogen , 220 x 100 / 46 = 478 kg Urea is required per hectare

Total area – 20.5 hectares

Hence, total requirement of Urea for 20.5 hectares = 478 x 20.5 = 9799 kg

Deduction for road/drain 4 % + 4 % for vacancy (due to death of tea plants) = 8 %

After deducting 8 % , total requirement of Urea = 9015 kg.

How Many Splits to Be Given in One Year ?

It is a one-million dollar question. It has been seen that many planters are confused in this regard but the process is very simple.

We know that in one split, the minimum quantity of Nitrogen which can be applied is 30 kg per hectare and the maximum is 100 kg per hectare. Hence, if by calculating the dose comes to 95 kg per hectare then only one split *i.e.* basal dose is enough. If the requirement is 267 kg per hectare , then 3 splits will be required.

So in this case , the number of splits for 220 kg Nitrogen per hectare will be 3 (three).

Time of Three Splits

In case of three splits, the application time should be as follows :

1St Split : Pre-monsoon , after receiving 5 cm of initial rainfall.

2nd Split: Just before the onset of monsoon.

3rd Split : Post monsoon, before the soil moisture dries up.

Split Percentage

1st Split – 40 %

2nd Split – 30 %

3rd Split – 30 %

So in this case , the split and the quantum of Urea for 20.5 hectares should be as follows :

1st Split – March-April – 9015 kg of Urea x 40 % = 3606 kg

2nd Split – June 1st week – 9015 kg of Urea x 30 % = 2704 kg

3rd Split- Mid September – 9015 kg of Urea x 30 % = 2704 kg

Requirement of Potassium

Since Potash status is alright and the section is under Light skiff, the ratio of N: K_2O should be 1 : 1. So the K_2O requirement in the present case is also 220 kg per hectare.

100 kg of Muriate of Potash (MOP) (Potassium chloride) contains 60 % K_2O. Hence for 220 kg K_2O, the requirement of MOP is 220 x 100 / 60 = 367 kg per hectare. For 20.5 hectares total requirement is 7523 kg.

Deduction for road/drain 4 % + 4 % for vacancy (due to death of tea plants) = 8 %

After deducting 8 % , total requirement of MOP = 6921 kg.

So in this case , the split and the quantum of MOP for 20.5 hectares should be as follows :

1st Split – March-April – 6921 kg of MOP x 40 % = 2768 kg

2nd Split – June 1st week – 6921 kg of MOP x 30 % = 2076 kg

3rd Split- Mid September – 6921 kg of MOP x 30 % = 2076 kg

Requirement of Phosphate

Generally it is said that Phosphorus should be applied in the form of Rock Phosphate or Single super phosphate in the pruned and Deep Skiff years. However, if fund permits, it is better to apply phosphate every year. The dose of P_2O_5 is 50 kg/ per hectare. In case of Rock Phosphate which contains 20% P_2O_5, the dose is 250 kg Rock Phosphate per hectare. In case of Single Super Phosphate , the dose of Single Super Phosphate is 312.5 kg per hectare.

Which Kind of Phosphate Should be Applied?

Single super phosphate(SSP) is a water soluble phosphate and Rock phosphate(RP) is an acid soluble phosphate. Hence, if the pH of the soil is less than 5.5 then Rock phosphate should be used and if it is more than 5.5 then SSP should be used. Rock phosphate is a slow releasing phosphate and found to be effective in the mature tea bushes.

Di Ammonium phosphate is given as foliar application in tea gardens @ 22 kg per hectare per year at 1 % concentration each time.

Basal or Split : Placement or Broadcast in Case of Phosphate Application

Phosphate is immobile in soil and leaching loss is almost zero. In few experiments, it has been noticed that placement of phosphatic fertilizers gives better result than broadcasting. Placement method is digging a small pit and covering with soil after putting the fertilizer.

In case of Light pruned and Medium pruned sections, it is better to apply phosphatic fertilizer in basal dose (once in a year) by placement method. It is better to dig a small pit in between two pruned bushes , put 30 grams of Rock phosphate and cover with soil.

In the skiffed sections , it is difficult to dig pits in between two plants because the top surface is covered up. Hence, in the skiffed sections, phosphatic fertilizers should be applied by broadcast method along with Nitrogen and Potash.

So in this case , the split and the quantum of Rock Phosphate for 20.5 hectares should be as follows :

Total requirement of Rock Phosphate is 250 kg per hectare (RP contains 20 % P_2O_5) X 20.5 hectares = 5125 kg. After deduction for vacancy + road/drains – 4% + 4 % (due to death of tea plants) = 8 % , it comes to 4715 kg for 20.5 hectares.

1st Split – March-April – 4715kg of RP x 40 % = 1886 kg

2nd Split – June 1st week – 4715 kg of RP x 30 % = 1415 kg

3rd Split- Mid September – 4715 kg of RP x 30 % = 1415 kg

Sulphur

Earlier the fertilizer programme used to made of NPK, but now the concept has changed, now it is NPKS. Now a days , along with NPK, Sulphur has also been made an essential component of fertilizer programme because Sulphur plays a great role in increasing flavor of tea . It contains three amino acids (cystine, cystiene and methionine) which serve as the precursor of VFC synthesis.

The dose of Sulphur is 40 kg Sulphur per hectare per year to be given in two splits, pre-monsoon and post- monsoon alongwith NPK. . Sulphur should not be applied during rainy season because it will evolve Hydrogen Sulphide which can be toxic to tea plants. When elemental Sulphur (which contains 80 % Sulphur) is applied in each split 25 kg of Elemental Sulphur should be applied per hectare.

Total requirement of Elemental Sulphur for 20.5 hectares is 1025 kg. After deduction for vacancy + road/drains – 4% + 4 % (due to death of tea plants) = 8 %, it comes to 943 kg for 20.5 hectares

So in this case, the split and the quantum of Elemental Sulphur for 20.5 hectares should be as follows:

1st Split – March-April – 943 kg of Elemental Sulphur x 50 % = 472 kg

2nd Split – June 1st week – No application

3rd Split- Mid September – 943 kg of Elemental Sulphur x 50 % = 472 kg

Final Calculation (for 20.5 Hectares) (in case of three splits)

	1ST SPLIT	2ND SPLIT	3RD SPLIT
	(40%)	(30%)	(30%)
Urea	3606 kg	2704 kg	2704 kg
MOP	2768 kg	2076 kg	2076 kg
Rock Phosphate	1886 kg	1415 kg	1415 kg
	1ST SPLIT	2ND SPLIT	3RD SPLIT
	(50%)	(0%)	(50%)
Elemental Sulphur	472 kg	0	472 kg
TOTAL FERTILIZER-	8732 kg	6195 kg	6667 kg

Now, it is required to calculate the fertilizer per plant. First, the number of plants per hectare to be calculated using the following formula:

Number of plants per hectare = (N x 10,000 square metres) / Y metres (X metres + Z metres)

(Where N= Number of hedges

X = Distance between hedge to hedge

Y = Distance between plant to plant

Z = Distance between row to row

In case single hedge N= 1 and in case of double hedge N = 2)

In this case

No. of plants per hectare = (1 x 10,000 m^2) / 60 cm (0 + 105 cm)

(here X=0 because there is only 1 hedge)

= (1 x 10,000 m^2) / 0.6 metre (0 + 1.05 metres)

= 15873 plants /hectare

Total number of plants after deduction for vacancy and roads/drains etc (total 8 %) per hectare = 14603. So, for 20.5 hectares, it comes to 299361 plants.

Fertilizer Per Plant

1ST SPLIT : 8732000 g / 299361 plants = 29 g / plant

2ND SPLIT: 6195000 g / 299361 plants = 21 g / plant

3RD SPLIT: 6667000 g / 299361 plants = 23 g / plant

To Increase Soil pH

Dolomite Doses (80-100 mesh size)

pH Range	Dose of Dolomite per hectare (in tones) Coarse textured soil	Fine textured soil	Time of application	Next Manuring
Below 4.5	2	3	January	At least after 8 weeks
4.5 to 4.65	1	1.5	January	

In the shade tree planting pit, addition of 1000 g of Dolomite can be very useful.

Application Method of Dolomite

Dolomite should be applied after forking of the interspaces. That is the reason that dolomite can not be applied during skiffed years. It should be applied after light or medium pruning when the interspaces are adequately exposed. A light irrigation after dolomite application really helps.

To Reduce pH

In many areas of Dooars, there is huge increase in pH because of Dolomite influx from the hills. To reduce pH, Sulphur/Iron Pyrites or Aluminium Sulphate is used.

Sulphur/Iron Pyrites or Aluminium Sulphate ?

Aluminium sulphate is easily available to the tea plants but Pyrites need microbial action for coming to the available form, hence, where the organic matter content in the soil is below 1 per cent , *i.e.* where the microbial population is less, it is suggested to apply Aluminium sulphate and not pyrites.

Sulphur in Planting Pits

Where the pH is too high, it is recommended to apply sulphur in the planting pit at the rates mentioned below:

Soil pH values	Sulphur per hole (g)
5.9-6.4	115
6.5-6.9	225
7.0-7.4	340
7.5 and more	450

(For 115 g of Sulphur , 450 g of Aluminium Sulphate is required)

In hut site areas where tea is not thriving well, Aluminium Sulphate @ 450 g per square metre placed on the ground every three months a year can be useful.

Soil pH and Avalability of Nutrient

Nitrogen: High pH – Availability increases

Low pH – Availability decreases

Phosphorus: In very acidic soil Phosphate gets bounded with aluminium, iron and calcium and forms aluminium phosphate, iron phosphate and calcium phosphate, hence, it becomes unavailable to the tea plants. Use of Phosphorus Solubilizing Microorganisms (PSM) or Phosphorus Solubilising Bacteria (PSB) like *Bacillus megatherium* and Vesicular Arbuscular Mycorrhizae (VAM) like *Rhizophagus thea* and *Protomyces thea* help in conversion to the available form.

Potassium, Calcium, Magnesium, Sulphur : With the decrease of pH, the availability of these nutrients decreases.

Aluminium, Iron, Manganese : At low pH also these nutrients are easily available to tea plants.

Copper, Zinc, Boron : High pH – Availability decreases

Low pH – Availability increases.

Molybdenum: In the acid soil Mo is precipitated with iron and aluminium. At higher pH , its availability increases.

Tea Board's Recommendation on the Range of pH and Nutrients

pH – 4.5 to 5.5

Nitrogen – 0.1 to 0.2 %

Organic Carbon – 1 to 2 %

Phosphorus (P_2O_5) – 10-20 ppm

Potash (K_2O) – 60-80 ppm

Sulphur – above 30 ppm

Tea Nutrition

The fertility of the soil depends on the microbial population in the soil more than the nutrient content. Tea being a perennial crop, the reserve of microbes in the soil is quite appreciable. However, the microbes are badly affected by the application of synthetic chemical etc. Nutrient reserve diminishes in the soil by means of plucking of leaves. It is reported that when 100 kg of green tea leaves are plucked, about 3.5 kg of Nitrogen is removed from the soil. Hence, replenishment is required by exogenous application of nutrients. Apart from broadcasting of fertilizers and nourishment of soil by using organic manures and bio fertilizers, it is necessary to spray nutrients , hormones as foliar application for inducing higher efficiency. In the tea plantation, micronutrients are given as foliar spray. Ground application of micro nutrients is not a practice in the tea industry.

Different kinds of foliar nutrients and growth promoters are used in tea to induce higher flush. Few propositions are furnished below :

Foliar Application of Nutrients

Foliar application means applying nutrients directly on the foliage of tea plants by sprayers after dissolving in water.

(1) Application of Urea is a general practice in tea. Almost every month Urea is applied @ 2 kg per drum (1% concentration).

(2) Di ammonium Phosphate (DAP) is applied , pre monsoon and post- monsoon at 1 % concentration totaling to about 22 kg per hectare per year.

(3) Zinc Sulphate is applied about 12.5 kg per hectare per year in many splits @ 1 % concentration. The efficacy of Zinc increasing with addition of Manganese @ 300, Boron @ 100 g and Alpha naptha Acetic Acid (NAA) @ 45 ml per drum.

(4) Spray of n-Triacontanol, a long chain fatty acid alcohol has proved to be very useful in tea.

(5) In many tea gardens gibberellic acid is applied once in a year , preferably in the month of May during dormancy. Gibberellic acid is applied @ 4 g per drum (200 litres of water). Since gibberellic acid is insoluble in water, 4 g of it is dissolved in 8 ml of Acetone and then transferred to a 200 litre-drum.

(6) Spray of a combination of Humic Acid and Fulvic acid formulation@ 1 % concentration is found to be very effective during lull (dormant) period and during drought.

(7) In many tea gardens, amino acids are sprayed @ 1 % concentration for vigorous growth of tea flush.

(8) Need based application of Magnesium Sulphate , 8 kg in a year @ 1% concentration is given.

(9) During water stress period, spray of NK foliar (1 % Urea and 1% Muriate of Potash) helps the tea plants to fight drought. Potassium regulates the function of stomata and induces stomatal resistance during dry period.

Role of *n*-triacontanol in Tea

- Promotes photosynthesis.
- Depresses phot-respiration.
- Keeps the cell turgid and induces succulence. Enhances water use efficiency.
- Activates efficient use of applied nutrients.
- Triacontanol is cost-effective and may increase yield upto 40 per cent.
- Triacontanol induces primary , secondary and tertiary growth.
- Tria increases polyphenol content.

NAMES OF A FEW GROWTH PROMOTERS , REGULATORS, GROWTH SUPRESSANT & ORGANIC MANURES AVAILABLE IN THE MARKET (2014)

	Trade Name	Company	Dose
Alpha naptha acetic acid (NAA)	Plantofix	Super Agro India	1ml/4.5 litres
	Planofix	Bayer Crop Science Ltd.	1 ml/4.5 litres
	Agrona	Aries Agro Ltd.	1 ml/4.5 litres water
Amino Acid	Fantac Plus	Coromandel International Ltd.	1.5 ml/litre
	Cheminos	Cheminova India Ltd.	1.5 ml/litre
	Aminocel Gold	Excel Crop Care Ltd.	1.5 ml/litre
	Samras	Multiplex Fertilizers Pvt. Ltd.	1 ml/litre
	Aminos	Rallis India Ltd.	1 ml/litre
	Amrit Sanjivan	Super Agro Ltd.	1 ml/litre
	Navras	Anshul Agro Chemicals	1 ml/litre
	Hydro pro	Aries Agro Ltd.	1ml/litre
Sea Weed Extract- Granules	Herbozyme GR	Excel Crop Care Ltd.	20 kg/hectare
	Biozyme GR	Biostadt India Ltd.	20 kg/hectare
	Cytozyme GR	Foliage Crop Solutions Pvt. Ltd.	15 kg/hectare
	Biovita GR	PI Industries	20 kg/hectare
	Superzyme GR	Super Agro Ltd.	20 kg/hectare
	Marino Gold	Aries Agro Ltd.	5 kg/hectare
Sea Weed Extract Liquid	Rapigrd	Isagro (Asia) Agrochemicals Pvt. Ltd.	1 ml/litre
	Rallizyme	Rallis India Ltd.	1 ml/litre
	Biovita	PI Industries Ltd.	1 ml/litre
	Hilgrow	Hindustan Insecticides Ltd.	1 ml/litre
	Biozyme	Biostadt India Ltd.	1 ml/litre
	Herbozyme	Excel Crop Care Ltd.	1 ml/litre
	Cytozyme	Foliage Crop Solutions Pvt. Ltd.	1 ml/litre
	Superzyme	Super Agro India	1 ml/litre
	Zyme Gold Plus	Godrej Agrovet Ltd	1 ml/litre
	EL++	Krishi Rasayan	1 ml/litre
	Marino	Aries Agro Ltd.	1ml/litre

Contd....

Table Contd....

	Trade Name	Company	Dose
Brassinolides	Bountee	Godrej Agrovet Ltd.	0.25 ml/litre
Herbal Amino Acid	Green Miracle	Indofil Industries Ltd.	1.5 ml/litre
	Isabion	Syngenta India Ltd.	1.5 ml/litre
	Siapton	Isagro (Asia) Agrochemicals Pvt. Ltd.	1.5 ml/litre
Humic Acid	Humifeast	Anshul Agro Chemicals	1 ml/litre
	Ecohume	Margo Biocontrols Pvt. Ltd.	1 ml/litre
	Earth	Amit Biotech	1.25 ml/litre
	Jivras	Multiplex Fertilizers Pvt. Ltd.	1.25 ml/litre
	Humicil	Dow Agrosciences India Pvt. Ltd.	1.25 ml/litre
Nitrobenzene + Folic Acid	Biostar	GSP Crop Science Pvt. Ltd.	2 ml/litre
Nitrobenzene 20%	Allwin Gold	Sree Ramcides Chemicals Pvt. Ltd.	2 ml/litre
	Jingold	Chemtura Chemicals India Pvt. Ltd.	2 ml/litre
Organic Manures	Annapurna	Karnataka Agrochemicals	475 kg/hectare
	Celrich	Excel Crop Care Ltd.	1000kg/hectare
	Ormacumin	Super Agro India	450 kg .hectare
	Vikas	Godrej Agrovet Ltd.	1000kg/hectare
	Orgafert	Aries Agro Ltd.	500 kg/hectare
n-Triacontanol 0.05%	Super photosynthate	S.B.Industries	0.5 ml/litre
	Miraculous	Super Agro India	0.5 ml/litre
	Miraculan	Dow Agrosciences India Pvt. Ltd.	0.5 ml/litre
n-Triacontanol 0.1%%	Vipul	Godrej Agrovet Ltd.	1 ml/litre
	Phytonol Gr + Mi	WBPPDCL	0.25ml/litre
	Miracle	Cheminova India Ltd.	0.5 ml/litre

There are many macro and micronutrient mixtures also available in the market.

6

VEGETATIVE PROPAGATION

Nursery Beds

Soil

pH should be 4.5to 4.8 (soil pH beyond 5 tends to promote excessive callusing at the expense of rooting). At higher level of pH, the roots of young plants die back (bitten off disease).

Sleeve Size

Lay flat	-	6" – 7" (15 – 17.5 cm)
Length	-	9" – 10" (22.5 – 25 cm)
Gauge	-	150

Bed Size

1.2 m X 30 m (2500 sleeves approx.)

Soil In The Sleeves

The lower ¾ portion of the sleeve to be filled with jungle top soil, rich in organic matter or soil : FYM :: 4:1 combination.

The top ¼ portion to be filled with rooting media comprising of sub soil.

Inputs

Following ingredients are added to the soil sleeve before planting of cuttings :

Trichoderma formulation– 1 -2 g + VAM (Vescicular Arbuscular Myccorhizae) – 5 g per sleeve

Single Super Phosphate

Add 1 kg of SSP per cubic metre of soil.

Drains

Between beds	:	Depth	-	45 to 60 cm
		Width	-	30 to 45 cm
Periphery	:	Depth	-	90 cm
		Width	-	60 cm

Materials Required for one Lakh Nursery Plants (In Case of Thatch Nursery)

Land – 0.2 hectare

Bamboo – 1200 (7 metres long)

Polythene Sleeves – 350 kg (150 gauge)

Soil – 70 truck load (250 MT) -2.5 kg per sleeve

Thatch – 2000 bundles

Coir String – 20 kg

Other Requirement for One Lakh Plants

Mother bush – 400 bushes

Man Power – 700 per one lakh plants per year

Chemicals – Acaricides – 0.8 litres ; Insecticides 2.5 litres; Fungicides – 2.0 litres

Soil Testing and Soil Treatment

Soil analysis for pH and eelworm is necessary. pH should be between 4.5 to 5.2. If the pH is high apply 50 ml of 2% Aluminium sulphate solution per sleeve. Equal quantity of water to be applied twice to leach down excess quantity of Aluminium Sulphate and obtain uniform pH. To make 2% solution of Aluminium Sulphate, dissolve 2 kg of chemical in 10 litres water and boil, then make the solution upto 100 litres.

Eelworm should not be more than 6 per 10 g of soil. If eelworm is more adopt the following method :

Calculate the quantity of soil required to fill 1000 polythene bags. Each time take this quantity of soil and mix nicely 1 kg of Carbofuran granules before filling the bags.

Shade

It is better to build overhead shade under which people can move freely. The height to be adjusted accordingly. The best thing would be to pt on Concrete pillars and Overhead structure and cover with netlon materials. The side walls should also be a netlon.

Arched Polythene Tent (Optional)

During Winter construct an arched transparent polythene tent, 1 – 8 m wide, over the sleeves. The bamboo frame should be 75 cm high. Polythene sheet of 400 gouge generally used. These sheets do not become opaque for 2 – 3 seasons. The sheet should not bear any leak.

The tent maintains high humidity inside and regulates temperature. During drought water the tent regularly from outside. The tent may be opened once in a fortnight (or ten days) and the bed may be watered inside.

Upkeep of Mother Bushes

Pinch off two leaves and a bud of the primaries to induce axillary buds, two to three weeks prior to harvesting the crop. This should be followed by foliar spray of 1% Urea solution and 2% aqueous solution of Zinc Sulphate (2 applications).

Time of Planting Cuttings

SPRING – April / May/June

AUTUMN : September/October/November

Time of Pruning/ Skiffing of Mother Bushes

Light Prune	Time Of Taking Scions
Mid October to 1st Week of November	End April to Early June
Mid November to 1st week of December	Mid May to End June
1st week of January to Early February	June – July

Deep Skiff

Early June	September / October
Early July	October/November

Preparation of Cuttings

The shoots from which the cuttings will be taken, should have the cut ends dipped in the water in a bucket and covered with fresh green plant materials. Green semi hard wood, one leaf cuttings consisting of One full leaf and an internode should be taken with a sharp knife. After each cut the cut end should be dipped in water. Immerse the separated cuttings in a bucket of water.

Planting of The Cuttings

The sleeves should be covered with black polythene sheet immediately after filling. In the evening, prior to putting out of the cutting, take out the cover apply 17 litres of water per square metre and again cover with polythene sheet.

Take the cuttings in the next morning. Punch small holes with a small stick, insert the cuttings gently. But before that dip the cutting (cut end) in 1% Zinc Sulphate solution. Do not dip the cuttings in Blitox solution. The node and the petiole should not be buried in the soil. The cutting should be planted in such a way that the leaves remain erect. The soil around the collar should be pressed hard to avoid air pockets. A light watering after planting of cuttings would be useful.

Fertilizer Application

When the plant puts forth 3 to 4 leaves apply YTD

(N:P:K: : 10:5:10) mixed with soil / Sand (1:9) @ 5 g per sleeve at 3 weeks interval.

Foliar Spray

On the plants having more than 4 leaves, spray :

First Round

DAP - 2000 g + Any Nursery Special formulation – 500 ml per 200 litres of water

Second Round

Urea - 2000 g + MOP-2000 g + Humic acid formulation– 200 ml per 200 litres of water.

Reasons for Poor Development of Plants in the Nursery

Symptoms	Reasons
1. High rate of mortality : brown stem	Dryness/ inadequate shade
2. High rate of mortality : soft bark	Wet bed / water logging**
3. Cutting live but no root	Weak cuttings/ deep shade/water logging high pH
4. Over callusing and weak root	High organic matter / high pH / water logging.
5. Poor growth : Yellow leaves	Nematode infestation
6. Poor growth : pale green leaves	Low nutrient / water logging
7. Poor growth : thick , hard leaves	Dry soil
8. Small sickle shaped leaves/ rosetting	Zinc deficiency of leaves

** **Reasons for water logging :**

(a) Poor quality of polythene sleeves

(b) Shallow drainage

(c) Too much of watering

Time for Development of Callus and Roots

Generally the cuttings take 4-6 weeks for callusing and 10-12 weeks for rooting under favourable conditions.

Shade Tree Nursery

Bed Preparation

It should be done during November-December. Deep ploughing followed by harrowing will be required. Loosen the soil up to a depth of 60 – 70 cm. On this soil, heap up dry jungle litters and burn.

SSP

After preparation of soil apply Single Super Phosphate @ 125 g per square metre of soil

(If directly planted on the beds)

Other Inputs

Well rotten cattle manure @ 25 MT and Dolomite (100 mesh) @ 2 MT per hectare.

(If directly planted on the beds)

Drains

Between beds : Depth – 60 cm ; Width – 45 cm

Periphery : Depth – 120 – 150 cm; Width – 60 cm

Size of Beds: Width 150 ; Length – As per availability of area.

Polythene Sleeves : Size – 30 cm lay flat ; Length – 60 cm ; Thickness – 300 gauge

Soil Mixture for Sleeves

Friable Soil – 4 parts ; Cattle manure – 1 part; SSP – 1kg per Cubic meter of soil; Dolomite – 0.5 kg per Cubic meter of soil

VAM – 5 g + *Trichoderma viride*– 1 -2 g per sleeve.

Sowing

Soak the seeds in water overnight followed by sun drying before sowing. Seeds with very hard seed coat can be treated with either hot water (80° C) or Concentrated H_2SO_4 solution. When treated with hot water, immerse seeds in water and heat upto 80° C and allow to cool overnight. If H_2SO_4 is used, soak for only 1-2 minutes followed by thorough washing in running water.

Time of Sowing Seeds

February to April

Method of Sowing

Directly on the Bed

Depth – 0.5 to 1.0 cm

Spacing – 20 cm

Interline spacing – 30 cm

Cover with thin layer of soil and water lightly.

Sleeves

Drill 2-3 seeds in each sleeve at a depth of 0.5 to 1.0 cm. Cover with thin layer of soil and water.

Pest Control

Soak the beds with insecticides after sowing. Pests to be controlled as per requirement.

7

YOUNG TEA MANAGEMENT

New Planting – Planting on the virgin land, where there was no tea plants earlier.

Replanting – When a tea plantation becomes economically unviable due to age or other reasons, replanting is done by complete removal of old plants and planting fresh lot of tea plants. Generally after 60 years of age, a tea plant does not remain economically viable, hence replanting is done.

Replacement – When a registered area of a tea plantation becomes unusable due to land slide or dolomite influx, the similar area of land in a virgin area is planted with young plants, this is known as replacement planting.

For Removal of Existing old Bushes

There are many methods employed for uprooting / removing old tea bushes.

Manual – The bushes are uprooted manually. In the hilly region where slope percentage is high it is difficult to uproot by using any machine, hence uprooting is done by hand. A worker can generally uproot 30-40 bushes per day.

No-tillage method – The collar is chopped off and on the chopped collar, 500 ml of 2,4-D solution is poured (1 kg 2,4-dicholoro phenoxy acetic acid in 75 litres of water). At some places, Tractor winch / Bulldozer / Crane – In the plains, the uprooting is done by any of these machines and these are found to be very effective.

As per Tea Board's regulation, in case of replanting, rehabilitation of the soil should be done for a minimum of 18 months after removal of old bushes.

Rehabilitation

Plants (cover crops) suitable for rehabilitation :

Tripsacum laxum (Guatemala grass) – The lopping of the plants adds to the humus of the soil.

Mimosa invisa (Thornless Mimosa)- Being leguminous in nature it has the capacity to fix atmospheric Nitrogen.

Pennisetum purpureum (Pusa Giant hybrid Napier grass)- The lopping of the plants adds to the humus of the soil.

Tripsacum laxum

Produces 100 tonnes of green matter per hectare. 10000 to 12000 cutting are required for one hectare. In highly acidic soil 2 tonnes of Dolomite (a mixture of Calcium carbonate and Magnesium carbonate) per hectare is broadcast 8-10 weeks before planting.

At the time of planting N: P: K : 10:5:10 @ 600 kg per hectare and 300 kg per hectare after lopping are applied.

Lopping – After the plant reaches 25-30 cm height, lopping is done at 8-10 weeks interval. Last lopping should be in October.

For one hectare plantation. 0.2 hectare nursery is required.

Pennisetum purpureum

Produces 200 tonnes of green matter per hectare. 10000 to 12000 cutting are required for one hectare. In highly acidic soil 2 tonnes of Dolomite (a mixture of Calcium carbonate and Magnesium carbonate) per hectare is broadcast 8-10 weeks before planting.

At the time of planting N: P: K : 10:5:10 @ 600 kg per hectare and 300 kg per hectare after lopping are applied.

First Lopping should be done 16-20 weeks after planting.

For one hectare plantation. 0.1 hectare nursery is required.

Mimosa invisa

3-4 kg seeds required per hectare. Rows should be 1.2 metres apart. Dolomite 2 tonnes per hectare should be broadcast 8-10 weeks before sowing. Single super phosphate @ 1 kg per 25 running metres to be applied in the rows.

Land Preparation

All the existing bushes and shade trees should be completely removed from the area to be planted. The roots should also be uprooted in case of normal uprooting method. In case of no- tillage replanting, the roots get decomposed by chemicals.

Topographic Survey : Before starting of drain digging, it is necessary to go for topographical survey which is also known as contour survey. From the topographical survey map, the plan should be drawn for proper drainage.

Drains : All category of drains viz main drain, sub main drain or collector drain and field drain or lateral drain should be dug before planting. Lateral drain and the row of tea plants should always be against the slope. The soil of the excavated drains should be spread properly so that there is no accumulation of soil along the drain side. The shape of the drain should be trapezoidal. In case of deeper drains, do not dig the whole depth at a time otherwise there can be huge soil erosion and the drains may collapse. It should be done in at least two phases. The dimension and spacing of drains should be as follows :

Spacing of Lateral Drain

Loamy sand - 18 metres

Sandy loam - 12 metres

Loam - 9 metres

Silt loam - 6 metres

Dimension of Lateral Drain

Drain	Drain depth	Top width	Bottom width
Lateral	105 cm	45 cm	15-20 cm
Sub-Main	120 cm	75-100 cm	30-50 cm
Main	150 cm	100-150cm	50-100 cm

The lateral drains should always be against the slope. At the meeting point of two drains, there should be a little curve so that soil erosion can be reduced.

Before digging the drains, criss- cross ploughing, criss-cross harrowing, criss-cross sub soiling (if required) and levelling should be done. All the stone particles should be removed from the land.

After drain construction, lay out for planting should be done.

Style of Planting and Spacing

Single hedge – Regular (rectangular) or staggered (triangular)

Double hedge - Regular (rectangular) or staggered (triangular)

Now-a-days, single hedge is preferred because in the double hedge, movement of workers, plucking and application of inputs become difficult.

Formula to calculate number of plants per hectare:

Number of plants per hectare = (N x 10,000 square metres) / Y metres (X metres + Z metres)

Where N= Number of hedges

X = Distance between hedge to hedge

Y = Distance between plant to plant

Z = Distance between row to row

In case single hedge N= 1 and in case of double hedge N = 2

The best spacing is 105 cm (between rows) x 60 cm (between plants) staggered single hedge.

Examples :

Single Hedge Plantation

Say the spacing is 105 cm (distance between row to row) X 60 cm (distance between plant to plant). So putting the values to the equation following result will emerge :

Number of plants per hectare = (N x 10,000 square metres) / Y metres (X metres + Z metres)

$= 1 \times 10{,}000\ m^2 /\ 60\ cm\ (0 + 105\ cm)$

(here X=0 because there is only 1 hedge)

$= 1 \times 10{,}000\ m^2 /\ 0.6\ metre\ (0 + 1.05\ metres)$

= 15873 plants /hectare

Double Hedge Plantation

Say the spacing is 105 cm (distance between hedge to hedge) X 75 cm (distance between row to row) X 60 cm (distance between plant to plant). So putting the values to the equation following result will emerge :

Number of plants per hectare

= (N x 10,000 square metres) / Y metres (X metres + Z metres)

$= 2 \times 10{,}000\ m^2 /\ 60\ cm\ (105\ cm + 75\ cm)$

$= 2 \times 10{,}000\ m^2 /\ 0.6\ metre\ (1.05\ metres + .75\ metres)$

$= 2 \times 10{,}000\ m^2 /\ 0.6\ metre\ (1.8\ metres)$

$= 2 \times 10{,}000\ m^2 /\ 1.08\ metres$

$= 20{,}000\ m^2 /\ 1.08\ metres$

= 18518 plants/hectare

Staking

With the help of plantation chains, the sticks (bamboo sticks) are fixed. The row of the tea plants should always be against the slope and along the lateral drains. During putting the sticks, marking should be made for the permanent shade trees also. In case of tea plants single stick and in case of shade trees double cross sticks should be put. After marking with the sticks, the pits should be dug manually or with the help of mechanical pit digger.

Pit Size

The depth should not be less than 45 cm and the diameter should not be less than 30 cm.

Inputs in the Planting Pits

Well rotten cow dung / Farm yard manure – 2 kg + Single super phosphate – 30 g. The cow dung or the FYM should be treated with carbofuran (in case of conventional garden) @ 2 g per kg of cow dung or *Trichoderma* powder (in case of bio-organic garden) @ 5g per kg of cow dung.

Soil Correction in Case of High pH

Soil pH value	Sulphur per pit (g)
5.9 to 6.4	115
6.5 to 6.9	225
7.0 to 7.4	340
7.5 and above	450

pH above 6 is in no way suitable for tea plantation.

Aluminium Sulphate

Places where organic matter content is less, it is better to apply Aluminium Sulphate @ 4.5 tonnes per hectare.

Soil Correction In Case Low pH

Soil pH value	Dolomite (tonnes per hectare)	
	Coarse textured soil	Fine textured soil
Below 4.5	2	3
4.5 to 4.65	1	1.5

Dolomite should be broadcast before digging of pits.

Specification of the Plants for Planting

Height – should be around 40-45 cm

Foliage – 12-16 leaves per plant

Stem Thickness – 0.5 to 0.8 cm

Planting Time

Planting should be done pre monsoon or post monsoon. In the area which is prone to water logging, planting should be done post monsoon and in the high lying droughty areas pre-monsoon planting can be useful. It is better to plant either early in the morning or evening. It is better not to plant when there is scorching sunlight.

Planting Method

While planting, it must be ensured that the axillary bud is not dipped inside the soil. After planting, proper ramming is required so that no air pocket is created. The accumulation of water in the air pockets can lead to death of the plants.

Mulching

Though water hyacinth and Guatemala grass are the best mulch materials, tea pruning litters, tea waste, *Citronella* , thatch and paddy straw can also be used for mulching. In many countries black plastic mulch is also used. For decomposition of the grasses, 100 kg of Ammonium Sulphate can be evenly spread on the mulch. Mulch should be kept at least 10 cm away from the plants to avoid heat damage.

Benefit of Mulching

- ☆ Mulching helps in conservation of soil moisture.
- ☆ It suppresses weed growth.
- ☆ It induces feeder root proliferation.
- ☆ It induces growth of microflora.
- ☆ It adds to the organic content of the soil.

Nutritional Values of Mulch Materials

MULCH	N %	P %	K %
Tea pruning litters	2.5	0.5	1.5
Tea Waste	3.6	0.6	2.3
Paddy straw	0.5	0.07	0.6
Thatch grass	0.5	0.1	0.5
Guatemala grass	1.6	0.4	1.3

Young Tea Manuring

Ytd mixture doses

$N:P_2O_5:K_2O::2:1:2$

Year of Planting	N (kg/ha)	YTD (kg/ha)	Method of Application
0 Year	20-40	200-400	Ring in 2-3 splits
+ 1 Year	80-100	800-1000	Ring in 4 splits
+ 2 Year	100-120	1000-1200	Ring in 4 splits
+3 Year	120-140	1200-1400	Ring in 4 splits
+4 Year	140-150	1400-1500	Strip in 2 splits

Note: Recommended by TRA

Young Tea Pruning

0 Year (the Year of planting) : In this year training of newly planted tea plants is done. Training is a method to promote lateral growth. One of the following methods is implemented about three months after planting, depending on the status, number of branches and health of the plants :

Very Weak Plants : Left as such and allowed to grow.

Single Stemmer With Majority of The Leaves Above 20 Cm from the Ground Level : In this case, "DE BUDDING" is carried out. In this method all the axillary buds and apical buds above 20 cm from the ground level are pinched off with the help of finger nails or forceps.

Single Stemmer With Plenty of Leaves Within 20 cm from the Ground Level : In this case "HAND LUNG PRUNING" or "THUMB BREAK" is done with the help of fingers. In this method, the leader stem is bent towards the Southern side at a height between 15 to 25 cm from the ground level. In this process, the vascular tissues (xylem and phloem) are disturbed in such a way that the free flow of solutes / translocation of solutes above the afore stated height is restricted, as a result lateral shoots develops from the lower nodes.

Plant With two or more branches : In this case the upright, central leader stem is removed with the help of a 7.5 cm bladed pruning knife. The method is known as "DE CENTERING" or "CENTRE OUT". No particular height can be assigned for this pruning because the middle stem should be cut at the junction at whatever height it is situated.

Plants Having Thick Stem and the First Lateral Situated at a Height more than 20 Cm : In this case "PARTIAL RING BARKING" is done. In this method, the bark at a height of 15 cm is cut in half circle manner.

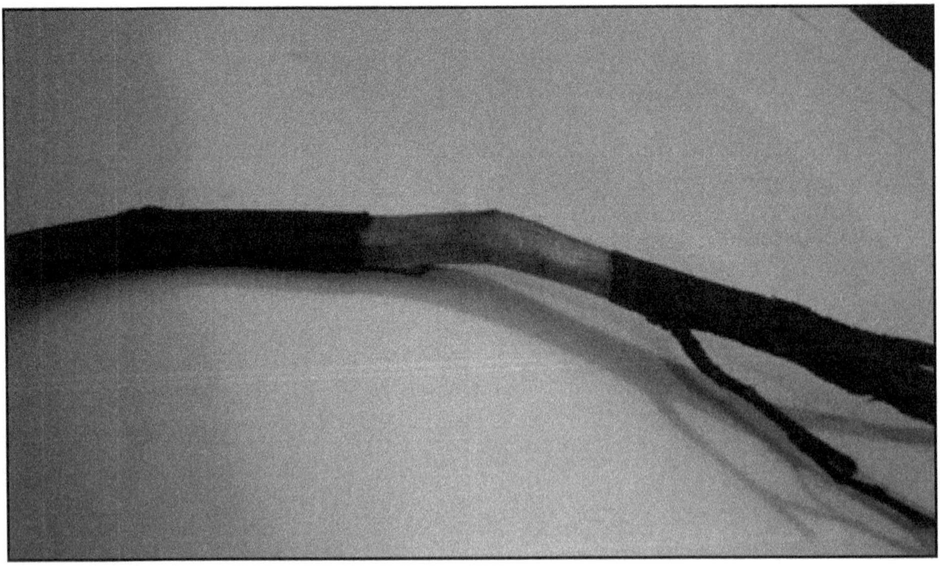

Ring barking

There are few more methods of Training of Young tea like "PEGGING" (in this method the branches are bent and pegged with a hook) , "HORMONAL OR CHEMICAL PRUNING" (in this case, growth suppressants like Cycocel, Paclobutrazol, Ethepon, Oxyfluorfen are sprayed on the decentred plants), but these methods are not popular.

+ 1 Year : In this year RCO (Re Centre Out) is conducted. The left over central leader stems are cut, the criss cross branches which tends to enter the centre , thin whippy branches are removed. The bush is skiffed at a height of 55 cm from the ground level.

+ 2 Year : In this year, the bush is pruned at 35 cm from the ground level. This is known as "FIRST FRAME FORMING PRUNING (FFP)". At the point of pruning the thickness of the stems should preferably be pencil thick. The cut marks should be as follows :

(a) Each cut mark should be round and small.

(b) The marks should not be oblique.

(c) The marks should not be like a tooth brush.

(d) All the marks should face the centre

(e) All the marks should be situated at the same height from the ground level.

(f) All the knots around the specified pruning height. To remove the knot, the pruning height may be adjusted according to necessity.

Following should be removed during the process by the process of "KNIFE CLEANING OUT (KCO)" by using a 7.5 cm long bladed pruning knife with hook type apical portion.

(i) All the criss cross branches.

(ii) All thin whippy branches.

(iii) One of the two branches at the close proximity. The weaker of the two branches should be removed.

(iv) All the hanging branches (Pankha daali).

Minimum number of branches should be removed during the First FFP. The bush should not become hollow centred.

A pruning stick, yellow painted with black mark at 55 cm level should be used. This stick should be kept in the middle portion of the bush. This stick should be used in each and every bush.

Breather : A breather should be kept in the middle and this should be removed when sprouting starts from the pruned branches.

+3 Year : In this no severe pruning is done. The bushes are given a Levelling of Skiff (LOS) (top skiff)

+ 4 Year : In this year, the bushes are pruned at about 40 cm from the ground level. **This is known as "FINAL FRAME FORMING PRUNING (FFP)".** In this year also Knife Cleaning out (KCO) is done.

Defoliation: The areas where infestation of red spider mites is a problem, it is better to remove all the leaves after Frame Forming Pruning.

Post FFP Treatments

(1) Immediately after FFP, spray any systemic fungicide @ 1 g per litre of water on the cut bushes.

(2) Knife Cleaning out to be followed within a week or so.

(3) The interspaces should be forked using fork hoe to loosen the soil so that Oxygen enters the soil pore spaces which in turn will promote Nitrification process.

(4) Organic manuring: In between two bushes, a pit should be dug and following ingredients should be inserted and covered with soil (placement method):

Well dried cow dung / Farm Yard Manure- 1 kg + Rock Phosphate – 30g per pit . FYM / Cow dung should be fortified using beneficial bacteria viz *Azotobacter, Azospirillum, Bacillus megatherium,* VAM etc. Now a days Organic manure fortifier are readily available in the market.

(5) Mulching should be done using water hyacinth or thatch grass or Guatemala grass or any other green matter. The mulching should be thick enough to keep the soil under cover for a considerable period. The mulch should be at least 10 cm away from the collar region of the tea plants.

8

DRAINAGE AND IRRIGATION

Good drainage involves quick and safe disposal of run-off and soil water in excess of its field capacity from the crop root gone without causing problems of water logging and soil erosion.

Benefits of Drainage

Improves root and shoot growth and leaf harvest through:

a. Roots go deep into the soil.

b. Reduces soil erosion and harmful effects of drought

c. Soil keeps warm in winter and cool in summer, which is favorable for root growth.

d. Improves soil aeration, soil structure, microbial activity, decay of organic matter and soil fertility

Problems Associated with Waterlogging

- ☆ Restricted out-fall.
- ☆ Run-off and seepage from higher catchment.
- ☆ Drain collapse.
- ☆ Sand boiling.
- ☆ Improper location and size of drains.
- ☆ Weed growth on drain bed and sides.
- ☆ Ponding due to uneven land surface.
- ☆ Subsoil impermeability.
- ☆ Unplanned roods, bunds, culverts etc.
- ☆ Wrong placement of culverts.
- ☆ Overflow from streams, rivers etc.

Areas suffering from water logging give low yields, have poor and patchy growth, poor response of fertilizers, increased effect of drought, heavy weed growth and higher incidence of pests and diseases.

Types of Drains

Subsidiary : These are laid down across the slope. (e.g. field, Lateral, Contour drains)

Primary : These are laid down along the slope. (e.g. Collector, Submain and Main drains)

Types of Drainage Systems

Random: Drains are dug out through natural depressions to drain out small, isolated low wet areas.

Herringbone: Where land slopes to middle from both sides, the collector drain is dug in the middle and the laterals join it from both sides.

Grid-iron: Laterals enter the callect6or drain only from one side.

Perimeter or Boundary Drain: Prevents entry of water from outside areas.

Interceptor Drain: It is dug near the upper edge of wet area, where slope meets the plot land, to intercept seepage water.

Open Drains

Advantages :

Initial cost low

Easy to construct and maintain.

Can carry large quantities of water.

Disadvantages:

Occupy large areas which can otherwise be put under plantation.

Requires culverts and bridges.

More difficult and expensive to maintain.

Pipe Drains

Advantages:

- ✰ Useful in sandy areas.
- ✰ No loss of planting area.
- ✰ Culverts and bridges are not required
- ✰ Subsequent maintenance costs practically nil

Disadvantages:

- ✰ Initial cost very high.
- ✰ Requires technical expertise to lay down these drains.

Causes of Drain Collapse

- ✰ Sedimentation – seen more if bed gradient is inadequate and not uniform.
- ✰ Improper location and alignment of drains.
- ✰ Improper junctions.
- ✰ Inadequate shape and size of drain.
- ✰ Excessive weed growth in drains.
- ✰ Blockage of outlet drain.

Trapezoidal shape of the drain

Drain Construction

- ☆ Dig drains during dry weather (NOV-MAR/APR).
- ☆ Dig main drain first, followed by submain and lateral drains.
- ☆ Start digging from outlet and proceed upgrade.
- ☆ Dig in stages.
- ☆ Spread the dug-out soil evenly over the area, or keep it a little away on lower side of the drain
- ☆ Avoid digging when soil is wet.
- ☆ Carry out repair of collapsed drains in dry weather.

Drain Maintenance

- ☆ Keep drain free of weeds, sediments etc.
- ☆ In sandy soils, grow deep rooting grass on drain sides.
- ☆ Avoid weeding on drain sides.
- ☆ Clean and re grade annually at least main/submain, drains.
- ☆ A good drain keeps flowing.

Culverts on Roads and Paths

- ☆ Use hume pipes of max. 1.2m dia for paths, and 2.4m dia for roads. RCC Slabs can be also used.
- ☆ Join two pipes firmly end to end with a collar.
- ☆ Bury the hume pipes at least 15cm into the drain bed, and put a minimum 45cm soil cover on top.
- ☆ Align the pipes with the drain, not with the road.

Spillway

This is required where vertical drop exceeds 100cm.

It may be made of bricks/stones in same way as slides in children's park.

Drain Specifications

Shape: Should be trapezoidal or half-trapezoidal.

Slope (Horizontal : Vertical)

Light soil (Loamy-sand)	0.5:1
Medium soil (Sandy Loam)	0.25:1
Heavy soil (silt loam, loam, clay loam)	0.25 : 1

Spacing of Field Drains in Flat Lands (Slope <3%) (Metres)

Soil type	Assam	Dooars/Terai
Clay loam	9	6
Silty loam	12	9
Loam	15	12
Sandy loam	18	15
Loamy sand	30	25

Depth vs Spacing for Medium Textures Soils

Depth	Spacing	
105 cm	18 m	Put the excavated soil on the lower side a little away from drain, or spread it evenly on ground.
120 cm	28 m	
135 cm	38 m	
150 cm	48 m	

Spacing on Sloppy Lands (Slope >3%)

In medium-textured soils, provide drains along contours as per following guideline:

Rainfall (cm)	Vertical interval (m)
$\geq$250	(% Slope +3) X 0.3048
251-400	(% Slope +2) X 0.3048
>400	(% Slope +1.75) X 0.3048

For other types of soils, make adjustments as follows:

Soil type	Adjustment (%)
Coarse (Sand, sandy, loam)	+20to +4
Med. Coarse (Loam, Silt. Loam)	+4 to -4
Med. Fine (Silt-loam, silty-Cloy-loam)	-4 to -12
Fine (Silty-clay, Clay)	-12 to -20

Design Specifications

	Lateral Drain	Submain drain	Main drain	Perimeter drain
Drain depth (cm)	105	120	150	150 – 180
Top width (cm)	45	75 – 100	100 – 150	150 – 250
Bottom width (cm)	15 – 20	30 – 50	50 – 100	100 – 150
Max.bed gradient (%)	0.25	0.15	0.10	0.25
Depth of flow (cm)	10	15	30	-
Max. flow velocity (cm/sec)	60 – 70	60 – 70	60 – 70	60 – 70

Drain Junctions

Drains should join together at an angle <90° and at bed level, with no out-fall.

Drain Location

Should be as straight as possible or with gradual curve.

Use existing drains as much as possible.

Put also a perimeter drain, if possible.

- If measured by use of float, multiply by a factor of 0.8

Selection of Drainage System

Flatland with slope up to 0.5% : Herring-bone system.

If outlet is restricted, 90 for pump drainage.

Land with mild slope of 0.5 – 1% : Grid-iron system.

Land with slope exceeding 1% : Provide interceptor

Drain across the main slope, and main drain down the main slope.

Undulating land with rolling topography:_ Grid-iron system of drains, with perimeter drain if required.

Land with basin-type topography: Requires pump drainage

Seepage from high lands : Deep interceptor drains across the main slope.

Water table fluctuates with rain fall : Improve surface and sub-surface drainage.

Soils gradually changing to more permeable strata with depth : Provide deep open/pipe drains.

Sandy loam and heavier soils : Provide suitably spaced open drains.

Sandy soils : Provide pipe drains

Few Definitions on Drainage

Catchment : Area bound by ridges and discharging excess water at one point.

Out fall or outlet : The point to which water is drained out.

Contour : Imaginary line drawn by joining points of same elevation with reference to certain datum line

Vertical drains : Wells dug through restrictive layer to drain water downward as in case of perched water table.

Relief Well : A shallow well carrying water under hydro-static pressure from a subsurface layer into a drain.

Sump : Reservoir created to facilitate operation of pump.

Sand Boiling : Rising of bed surface due to pressure exerted by seepage water on light textured soil.

Making of Drainage Schemes

Determine the size of area to be drained.

Carry out topographic survey and prepare contour map.

Contour interval : 30 cm for flat land

100 cm for steep topography

Scale for ≤120 ha – 1:1000, 121 to 240 ha – 1:2000,

>240 ha – 1: 4000

Based on topography, select type of drainage system and mark location of drains.

Study soil properties and decide about shape, side slope and bed gradient of drains.

Study quantum and pattern of RF distribution

Outlet availability, its elevation and distance.

Size and location of bridges, culverts, roads etc.

Average and high flood levels of drainage outlets.

Drain Stabilty

Level of under-ground water table and its fluctuations.

If required, LS profile of existing main/sub main drains.

For each catchment, develop main drains following natural water course starting from out6fall. Avoid sharp bends.

Provide collector drains along the slope, one for each sub-catchment.

Mark lateral drains across the slope in flat lands, and along contours in slope in areas.

Provide bed gradient from the highest point to the lowest point of outfall.

Master Row Planting

Location of master row is done on following principle:

- ☆ If a pair of graded contour drains converge in their direction of flow, the master row is made parallel to the top or higher contour drain.
- ☆ If a pair of graded contour drains diverge in direction of their flow, the master row is made parallel to lower contour drain.

Calculation of size of Main/submain Drains

Volume of Water Discharged

Q = 0.0028 C;A or Q = DCX A, where

Q = vol. of water discharged (m3/Sec)

C = run – off Coefficient (0.3 – 0.5, higher for heavy soil)

I = Intensity of R.F. (mm/hr)

A = Catchment area (ha)

DC = Drainage coefficient or depth of water to be removed (mm/sec)

Bottom width of Drain

A = bd + zd2, where

A = Area of water flow (sq.m.)

B = Bottom width (m)

D = Design depth of flow in drain (m)

Z = Drain side slope (horizontal/vertical)

Area of Water Flow

A = Q/V, where

A = Cross section of drain water flow (sq.m.)

Q = Volume of water discharged (cu.m./sec)

V = Velocity of water flow (m/sec)

Few Tips on Irrigation

Irrigation Schedule

For Unpruned Areas

	Irrigation	Frequency
Nov	2"	once a month
Dec	2"	once a month
Jan	1"	once a month
Feb	1"	once a month
Mar	2"	once a month
Apr	4"	2" fortnightly
May	6"	2" every 10 days

For LP and DS Areas

Jan	1"	once a month
Feb	2"	once a month
Mar	2"	once a month
Apr	2"	once a month
May	4"	2" fortnightly

NOTE : 25 mm or 1" rain equals 250 tons / hectare ; say 10000 l pump running for 25 hours
(Courtesy Tea Research Association)

Few Equations on Sprinkler Irrigation System

* Required discharge from an individual sprinkler (litres/second) $= \dfrac{\text{SI x Sm x 1}}{3600}$

SI = Spacing of sprinklers on the lateral line (mitres)

Sm = spacing of laterals on a main line (metres)

I = rate of water application (mm/hour)

Total pump discharge = N X Q

N = No. of sprinklers run at a time.

Q = discharge from each sprinkler (litres/second)

Discharge capacity of sprinkler system (litres/ second) = $\frac{2.78 \times A \times d}{F \times H \times E}$

A = Area irrigated

d = gross depth of each irrigation (mm)

F = No. of days allowed for completion of one irrigation

H = No. of actual operating hours per day

E = Irrigation efficiency (say 80%)

Performance of Twin Nozzle Sprinklers

Nozzle Diameter (mm)	Pressure psi	Diameter in ft.	Discharge in gpm	precipitation 24x24 m	inches 24x30m	hour 30x30m	setting 30x 36m
10 x 7	40	152	31.6	0.59	-	-	-
	45	154	33.5	0.62	0.50	-	-
	50	158	35.8	0.67	0.53	-	-
	55	163	37.5	0.70	0.56	-	-
	60	168	39.3	0.73	0.59	0.47	-
	70	178	42.5	0.79	0.63	0.51	-
	80	188	45.0	0.84	0.67	0.54	-
12 x 7	40	159	40.6	0.76	0.60	-	-
	45	163	43.0	0.80	0.64	0.51	-
	50	168	45.6	0.85	0.68	0.54	-
	55	173	48.0	-	0.72	0.57	-
	60	178	50.3	-	0.75	0.60	-
	70	188	54.7	-	0.82	0.65	0.54
	80	198	58.1	-	0.87	0.69	0.58
14 x 7	45	171	57.5	-	-	0.69	-
	50	176	61.0	-	-	0.73	-
	55	182	64.3	-	-	0.77	-
	60	187	66.9	-	-	0.80	0.66
	70	197	72.4	-	-	0.86	0.72
	80	207	77.2	-	-	0.92	0.77

Contd....

Table Contd....

16 x 7	45	177	69.0	-	-	-	0.69
	50	182	73.2	-	-	-	0.73
	55	187	76.8	-	-	-	0.76
	60	192	79.9	-	-	-	0.79
	70	203	86.6	-	-	-	0.86
	80	213	92.2	-	-	-	0.92

Area Coverage Per Sprinkler Per Setting

24 m by 24 m	=	80' by 80'	=	6400 sq.ft	=	0.147 acre
24 m by 30 m	=	80' by 100'	=	8000 sq.ft	=	0.184 acre
30 m by 30 m						

Tensiometer

To know about when to start irrigation or when to stop irrigation, installation of tensiometers would be essential.

Tensiometer Installation

A pipe which is 2-3 mm less in diameter than the tensiometer tube diameter is used .

The metal pipe is hammered into the soil to a depth of about 10 cm less than the size of tensiometer and the pipe is pulled out with soil leaving a vertical hole behind.

A little water is poured into the hole to make the bottom soft to accept the tensiometer.

After the water sinks into the soil , gently the tensiometer is pushed into the core vertically so that the porous cup firmly gets into the soft soil below the hole maintaining a close contact with the soil.

The soil around the tensiometer body is then tamped at the surface to seal around the body tube to prevent surface water from entering around the tensiometer body.

After installation , several hours required before the tensiometer reads the correct soil suction value. Protect the tensiometer from any damage by people and animals by constructing a box on the tensiometers.

Tensiometer Readings

0 - 0.10 : it shows soil saturation.

0.1 – 0.2 : soil moisture in field capacity. The spriklers may be turned off at this point.

0.2 – 0.5 : Good availability of soil moisture to plants. Irrigation not required.

0.5 – 0.6 : It is the usual range for starting irrigation.

Higher than 0.6 : Moisture stress range.

9

DROUGHT MANAGEMENT IN TEA

Drought has become a way of life in the tea industry of North East India. Every year the dry spell intensifies from March to May, resulting in huge crop loss, wilting of tea plants and eventually death of plants in stony and sandy patches of tea gardens. Since changing of climate is not in the hands of tea planters, it is pertinent to train tea bushes to adopt this new phenomenon and gear up for fighting battle with drought.

Impact of Drought on Tea

- There is a general reduction in size of the plants subject to prolonged water stress.
- Reduced cell turgor is the most important reason for reduced plant size.
- Changes in structure of proteins caused by decrease in water potential effect enzyme activity.
- Osmotic potential is important since the protoplasm must be in osmotic equilibrium with the sources. Water stress decreases O.P. Small decrease in osmotic potential may be accompanied by a large decrease in turgor pressure and water potential.
- Decrease in starch content.
- Decrease in soluble protein.
- Inhibits auxin formation.
- Cytokinin activity reduced.
- Translocation reduces.
- Proline (Pynolidine – 2- Carbosylic acid) increased.

Reports on Water Stress

- ❖ Reduction in water content usually reduces the rate of photosynthesis both because dehydrated protoplasm is less effective and because stomatal
- ❖ Decrease in photosynthesis at high values of stress may be due to stomatal closure.
- ❖ Inhibition of photosynthesis by water stress may be attributed to both stomatal and non- stomatal factors depending on particular environmental conditions.
- ❖ Tea bushes are benefited by high atmospheric humidity. He added that low humidity affects the physiological activities adversely by influencing the cell sap concentration.
- ❖ Water stress in the plant directly reduces the rate of photosynthesis because the dehydrated protoplasm has a lower photosynthetic capacity.
- ❖ High atmospheric stress is expected to reduce photosynthesis probably causing rapid evaporation from guard cell and causing stomata to close.
- ❖ Decreasing water potential in leaves is known to reduce photosynthesis.
- ❖ Reduction in photosynthetic rate is the result of three effects :

  i) Hydroactive closing of stomata, bringing about a reduced CO_2 supply.

  ii) Water stress in cytoplasmic ultrastructure affecting enzyme activity.

  iii) Dehydration of cuticle, epidermal walls and cell membranes reducing their avidity for and permeability to CO_2.

Role of Water Potential

- ❖ Photosynthesis decreases with low leaf water potential.
- ❖ Drought tolerant clones consistently exhibit higher water potential.
- ❖ Decreasing water potential in leaves is known to reduce photosynthesis.
- ❖ Low water potential causes metabolic imbalances within plants, cell elongation being one of the most sensitive portion .

Role of Stomatal Functions

- ❖ Stomatal conductance reduces due to water stress.
- ❖ Higher stomatal resistance affects gaseous exchange and consequently the rate of photosynthesis.
- ❖ Stomatal resistance is the principal limitation to photosynthetic CO_2.
- ❖ Stomatal resistance increases under water stress condition.

What to be done to Combat Drought ?

(A) Planting Material

In drought prone area it is pertinent to select clonal variety and seed plants which have drought tolerance potential. In high lands, where water table is too low, rainfall is scanty and rain fed areas , it is better to go for seed variety rather than clone because clonal plants have shallow adventitious roots but a seed plant has a strong tap root system which can pull water from the lower hamper of soil profile.

(B) Higher Percentage of Pruning

In a drought prone area, the percentage of Light Prune should be on higher side because a pruned bush suffers less in drought due to late development of leaves. Remember, the transpiration loss of water is much more than evaporation from soil.

(C) Irrigation

Selective irrigation with sprinkler irrigation system can be useful for the plants.

(D) Potassium Plays a Great Role

(1) Potassium increases water use efficiency and turgidity, regulates stomatal function.

(2) Various experiments all over the world have proved that Potassium imparts drought tolerance. Guard cells of higher plants control transpirational water loss and gas exchange for photosynthesis by opening and closing of pores in the epidermis of the leaf.

(3) Potassium stabilizes the emulsions of colloidal particles that keep the protoplasm hydrated and cellular membrane permeable.

(4) 4-6 applications of Urea and MOP each at 1 % at about monthly intervals during the dry period are recommended in young/mature tea to maintain turgidity of cells, improve water use efficiency and to reduce banjhi shoot formation. The study indicates that under moisture stress, increased survival, improved dry matter production and yield of tea plants were due to improved K bio availability following K application. Larger amount of K is required under moisture stress compared to normal conditions .

(5) Potassium is recommended at the N: K_2O ratio of 1:1 irrespective of source of Nitrogen and soil test values (Handbook of Tea Culture, UPASI TRF). In the pruned year, N:K_2O ratio should be maintained depending on the height of pruning. For height below 45 cm, N:K_2O ratio should be maintained as 1:2, between 45cm and 60 cm as 2 : 3 and above 60 cm as 1 : 1.

(F) Mulching in Young Tea

Young tea is the mostly affected during drought because exposure of surface soil is much more in young tea compared to matured tea. Hence to reduce direct evaporation from soil and protect the soil from direct exposure to scorching sunshine , it is necessary to cover the soil with mulch materials like water hyacinth or any green matter or even plastic mulch. Buffering of temperature by mulch can help in preserving more water in soil, hence activate uptake of nutrients from soil.

(G) Role of Shade Trees

Shade trees play a great role in reducing the water stress. Shade trees not only cut off detrimental solar insolation but also prevents evapotranspiration. Shade trees

Water hyacinth

in tea gardens are mostly deciduous , hence, those shade trees should be planted in droughty areas which can retain leaves during March – April.

(H) Organic Matter Reduces Water Stress

Organic matter content in soil has higher water holding capacity .Hence application of Vermicompost or other compost really helps.

10

BIO-ORGANIC FARMING/BIOLOGICAL CONTROL OF PESTS

The term 'Bio-Organic Farming' is used for the plantations where no synthetic chemical pesticides and inorganic fertilizers are used. It is better to term as 'Bio-Organic Farming' rather than only ' Organic Farming' because most of the synthetic pesticides are organic in nature. Although very low in number, there are many tea gardens where the concept of bio-organic farming is strictly followed.

Synthetic pesticides and inorganic fertilizers induce environmental pollution and health hazards. Although through various seminars , various publications it has been endeavoured to generate awareness among people about menace of synthetic chemicals, the level of reduction in chemical use is still insignificant. After the publication of the book 'Silent Spring' by Rachel Carlson , there has been a growing consciousness among the planting community to reduce the use of conventional synthetic chemical pesticides and keep the residue level under control. At one point of time, it was almost impossible to replace conventional pesticides and fertilizers because alternatives were hardly available. But today, host of alternatives are available in the market. This revolutionary change in attitude of the manufacturing companies has ushered in a new era in easy conversion of conventional tea garden to a bio organic one, without any/ much set back in productivity of tea.

Approximately one third of world food production is lost due to pests. If it were not the pesticides another one third would be lost. Nobel laureate Borlaug said in 1979 'we have to choose between feeding human beings or feeding pests'. A survey by National Council of Applied Economic Research has revealed that 'Every one rupee spent in plant protection would protect crop output worth rupees four , which would have been lost to pests'. While it is true that to increase crop yield pesticide has a great role to play but the application needs to be judicious . Excessive use of pesticides has become hazardous to human health as reported by many author.

Heavy use of chemical pesticides has taken serious turn in Thailand on the 'Hmong Farmers' health . It was observed that in Laguna and Nueva Eeija area of Philippines where the farmer use monocrotophos had health problem twice as compared to the farmers of Lucban and Quezon province who rarely used pesticides. It was reported that in 1989-90 more than 665 people died in India because of Pesticide poisoning and thousands of cases of crippling from pesticide have been recorded. With the growing consumption of pesticides the resistance of the insects against pesticides also increased. In 1934 there were 10 insects were known to be resistant against pesticides whereas by 1980 the figure had risen to 432. Pesticides were also found to affect the population of *Azotobacter* , *Rhizobium* etc. It reduces the number of nodules and affects mycorrhizal symbiosis. Owing to excessive use of pesticides alternation in the nitrogen balance occurs because pesticide affects nitrifying bacteria *viz. Nitrosomonas* and *Nitrobacter*. The list of adverse effect of synthetic pesticides in quite long and this emphasizes the need to search for other alternative method which would be effective in controlling the pests which causes economic injury to the crop and at the same time reduces the risk of health hazards.

There should be two distinct sets of practices in a bio-organic tea garden : (i) Biological Pest Control (ii) Biological Nutrient Management .- delete

Demerits of Synthetic Pesticides

(i) Most of the pesticide residues affect the central nervous system, respiratory system and gastro-intestinal system of human beings .

(ii) Depression, insomnia, hyper reflexia of human being are also caused by pesticides.

(iii) The chemicals affect ground water resources through seepage.

Merits of Bio-organic Culture

(i) It does not pollute the soil and ground water with chemical residues.

(ii) It increases the biological diversity of plants and animal.

(iii) It prevents leaching of minerals in the soil.

(iv) It utilizes natural equilibrium for crop protection.

Do not Jump to Bio-organic Overnight

There can be a considerable reduction in productivity , if suddenly a tea garden switches over to bio organic form from a conventional form. There can be two distinct methods for such conversion :

(i) Transform to bio organic through the pathways of Integrated Pest Management (IPM) and Integrated Nutrient Management (INM).

(ii) Convert one third of the tea garden into bio organic and then proceed in phasic manner.

Integration of Various Methods of Pest Management Systems

1. Plant Resistance (Varietal Control)

(Genetic make-up, tolerance use of mixed varieties)

In this case the resistance and susceptibility of a tea clone or seed stock in respect of any pest is taken into consideration. For example TV 23, Phoobshering 312 are susceptible to red spider mites; Teenali 17/1/54 is somewhat resistant to red spider mites. Hence in an area where infestation of mites is a problem better to avoid planting of TV 23 and Phoobshering 312. Tocklai Vegetative 1 (TV 1) is susceptible to Tea mosquito bug, hence where Tea mosquito bug infestation is prevalent, it is better not to plant TV 1. Tocklai Vegetative 9 (TV 9) is resistant to Violet root rot disease caused by *Sphaerostilbe repens*, hence for the waterlogged sections, planting of TV 9 should be considered as a good clone.

2. Cultural Control

(Destruction of alternate hosts, barrier crop, raising of shade trees, irrigation & fertilizer practice, mulching.)

Most of the insects / mites need alternate host as weeds for completion of their life cycles, hence weed control is one of the best cultural control to keep pest population under control. In case of red spider mites, dust deposition on the leaves serves as the breeding ground for the mites, hence raising barrier hedge like *Adhatoda vasica, Duranta plumeiri* and *Lantana camara* can be planted on the boundary to prevent entry of dust. Even a row of tea plants along the roads can be left un plucked to take the shape of a hedge to prevent dust entry.

There are two categories of pests found in the tea garden. One is positively thermo (temperature) and photo (light) tropic, which means, if the light and temperature are more , these kind of organisms increases its population massively due to congenial weather conditions example red spider mites, thrips etc. Another type is negatively photo and thermo tropic, which means its population will increase where temperature is low and solar radiation is less example Tea mosquito bug, *Exobasidium vexans* (the causal organism of blister blight) etc. Hence, in case of the first type, adequate shade trees, which retains maximum foliage during dry period / drought should be there. There should be frequent irrigation to create a micro climate in the phyllospheric region to control pests.

In young tea, mulching i.e. covering of the barren surface with green mulches like water hyacinth, Guatemala grasses , thatch etc can be useful in maintaining a congenial agro-ecological condition, which will keep the pests under control.

3. Biological Control

(Predators, parasitoids, pathogens, microbial pesticides e.g. *Bacillus thuringiensis;* Botanical antifeedant & repellants.)

Microbial pesticides, both entomo-pathogens and phyto - pathogens are now easily available in the market. Use of those against selective pests can be useful in controlling pests.

Plant extracts play a great role in controlling pests. Many weeds are used now-a-days for extracting juices, which act as anti feedant and repellant.

4. Interference Method

(Pheromone traps, light traps, Sterile male Technique, Insect Growth Regulators e.g. diflubenzuron)

Although Interference method is still not very popular, it has considerable role in controlling pests. Different types of pheromone traps and light traps are available in the market now-a-days and many tea gardens have started using these devices. Insect growth regulators are also manufactured by many companies.

5. Chemical Control (Chemical pesticides)

An elaborate list of chemical pesticides have been included in this book, with the chemical names, trade name, name of the manufacturing company, doses and target pests which can be used by the tea planters as a ready reckoner.

Appropriate Management of Pest Control Techniques

Pest control programme will only be successful if appropriate management is done. It is necessary to identify the pests before applying any pesticide. Knowledge on the location of the particular pest in a plant , their life cycles, their damaging stages and period is of utmost importance. Spray techniques and spray schedule can only be properly prepared if the planter has a thorough knowledge about a pest. For examples are given below:

Target Pests

(i) To control thrips , jassids and scarlet mites, it is necessary to spray the under surface of the leaf, because these insects and mites remain on the ventral surface of the leaves.

(ii) To control tea mosquito bug, it is necessary to spray during early morning and evening because during heavy sunshine tea mosquito bugs escapes to the drains and weeds. It is also necessary to go for barrier spray in case of tea mosquito bug. Barrier spray means, the boundary should be sprayed first using any fumigant so that the pest does not migrate to the adjacent section of tea. Tea mosquito bug is very fast in migrating.

(iii) To control looper caterpillars, it is necessary to spray the shade trees which serves as the alternate hosts of the caterpillars using foot sprayers or Bak Pak sprayers with BAN nozzle.

(iv) To control cockchafer grub, it is better to spray during evening because, they are more active during evenings.

Proper Pesticide

Right kind of pesticides should be used to control pests. In case of wrong selection of pesticide, there will not only be wastage of money due to pesticide but it will also inflict crop loss due to delay in controlling.

Pest Mapping and Pest Surveillance

Each tea garden should prepare a pest map for each tea section. The history of the section as far as pest infestation is concerned should be thoroughly studies. Data on agro-ecological condition, soil features, physiography, shade status, planting material, pest infestation type and time should be collected to prepare a section –wise pest map. This map will help the tea estates to plan their replanting programme, pest control strategy etc.

Every week a thorough study of the infestation status of pest in various section should be studied and a Pest Surveillance Weekly chart should be prepared to draw pesticide spray programme. This will serve as a radar for preparing pest control methodology.

Correct Chemical and Correct Doses

Right kind of pesticides should be chosen according to the species of the pest and intensity of infestation. Doses of the pesticides should be appropriate so that it reaches lethal dose, otherwise it may not be effective. Under dose can be very dangerous and it can increase the immunity of the pests.

There should be two distinct sets of practices in a bio-organic tea garden : (i) Biological Pest Control (ii) Biological Nutrient Management.

Biological Pest Control

Protocols in Complete Bio-organic Sections

Control of Red Spider and Other Mites

Cultural Control

REGULATED SHADE : Mites are positively thermo and phototropic organisms. Too much infiltration of solar radiation and too much increase in temperature induce population explosion of mites. Hence it is necessary to cut off detrimental solar insolation and create a micro climate in the phyllospheric region conducive for the plants, by planting suitable shade trees. The species of shade trees should be selected taking into account the leaf fall period. Almost all the shade trees are leguminous , hence deciduous and shed leaves once in a year. It has to be seen that the leaf fall period of the chosen shade tree should not coincide with droughty period of the garden.

Shade trees have other important functions too:

(i) Loppings, fallen leaves and twigs serve as humus

(ii) Most of the shade trees being leguminous, have Nitrogen fixing capacity

(iii) Creates a micro climate which provides soothing effect to the workers. Pluckers' productivity and workers' efficiency increase significantly under regulated shade

(iv) Strong root system of the shade trees, breaks hard pan in the soil and paves the way for tea roots to proliferate. It also increases aeration inside the pore spaces

(v) Shade trees act as wind barrier and protects the plantation from damage

(vi) Thinning outs and loppings serve as fuel wood

(vii) Controls many pests by buffering temperature

Total Control of Weeds : Weeds serve as alternate hosts for most of the insects and mites. These organisms (pests) complete their life-cycles in two different species of plants. If weed is controlled, life cycle of the organisms is disturbed and reduction in population occurs.

Adhatoda vasica

***Duranta plumeiri* hedge**

Prevention of Dust : Dust deposition on the leaves creates congenial substratum for laying of eggs of the mites. Hence it is necessary to prevent dust, especially in the roadside blocks, by raising boundary hedges.A row of tea bushes can be left unplucked after pruning or hedge plants like *Adhatoda vasica* or *Duranta plumeiri* or *Lantana camera* can be grown.

Deployment of Workers : Workers often serves as vectors (carrier) of mites. Through their bodies, plucking baskets etc mites are transmitted to unaffected sections. Hence deployment of workers should be from uninfested sections to the infested sections. Cattle also often serve as vectors. Hence, entry of cattle should be prevented.

Irrigation : When the temperature increases considerably and moisture stress

(vapour pressure deficit) is high, it is necessary to go for irrigation to induce humidity. Irrigation lowers the temperature thereby reduces the chance of mite infestation.

Biological Control

During Jan - Feb

Spray at least two rounds of Sulphur – Neem combination with thorough drenching of bushes in the skiffed sections. BIO-PESTICIDES

Paecilomyces fumosoroseus and *Verticillluim lecanii* in alternative rounds can be sprayed when population build up is nominal. Once , it is allowed to cross Economic Threshold Level (ETL), it will be difficult to control mites biologically. So, Pest

Surveillance Technique is of immense importance in monitoring pest build up. It should be implemented meticulously to nip the pests in the bud stage.

Control of Sucking Pests

(Tea mosquito Bugs , Thrips, Jassids,Aphids etc.)

Cultural Control

- To control sucking pests , it is necessary to keep the plantation weed free.
- To control thrips, it is necessary to give a caustic /lime wash to the bush frame after pruning in the following ratio: Lime 1 kg + Caustic soda or washing soda @ 3 kg in 50 litres of water.
- To control tea mosquito bugs, black plucking, slashing should be done if required.

Biological Control

Spray *Beauveria bassiana*. While spraying against tea mosquito bugs, it would be necessary to go for barrier spray.

Control of Caterpillars

Cultural Control

To control caterpillars , it is necessary to hand collect the pupae of the caterpillars from the bush and soil.

Biological Control

Bacillus thuringiensis and NPV. In case of red slug caterpillars, mid hamper of the bushes should be targeted.

Other Pests

To control nematodes , ground application of *Paecilomyces lilacinus,* to control cockchafer and termites , *Metarhizium anisopliae* can be useful.

To control diseases caused by fungus and algae, spray of *Trichoderma viride* and *Pseudomonas flurescence* can be very useful. Cu- oxychloride is also allowed in organic tea gardens to control fungal and algal pathogens.

To Control Red Rust

Red rust disease is a secondary disease. Pre disposing factors for Red rust attack are (i) Low Potassium content in the soil (ii) Too dry condition and iii) waterlogged condition. Hence red rust disease which is caused by *Cephaleuros parasiticus* can be controlled by regulated shade, higher dose of Potassium and deepening of drains.

Spray of Bio-control Agents

Spray while there is sufficient moisture in the atmosphere and soil.

Use a sticker for better result.

Spray while there is minimal sunshine

Wash the spray machines and containers before spray

There should not be any spray or broadcast of fungicide or bactericide before & after seven days' of spray .

Post Pruning

To control *Poria* attack, immediately after , pruning, apply *Trichoderma* paste on the bigger cut marks and spray *Trichoderma* formulation@ 1 ml or 5 g per litre on the bush frame.

Spray of Bio-control Agents

Spray while there is sufficient moisture in the atmosphere & soil

Use a sticker for better result

Spray while there is minimal sunshine

Wash the spray machines and containers before spray

There should not be any spray or broadcast of fungicide or bactericide before & after seven days' of spray .

Addition of Neem Formulation: With the bio pesticides, if neem product (*Azadirachta indica*) at recommended doses is added , efficacy can go up quite considerably and control may be prolonged.

Post Pruning

To control *Poria* attack, immediately after , pruning, apply *Trichoderma* paste on the bigger cut marks and spray *Trichoderma* formulation @ 1 ml or 5 g per litre on the bush frame.

Predators / Prasites / Parasitoides

Predator

A predator is an organism that captures and eats another (the prey). This act is called predation.

Features of Predators

- They are usually larger than their prey, or overwhelm their prey by attacking in large numbers like ants.
- Most do not have specific prey and feed on a wide range of animals which, in some cases, they consume in large numbers. Lady beetles for example, can devour several hundred aphids in a day.
- Some are specific, targeting a particular prey group.
- Death to the victim is usually immediately after capture.
- Adults or young may be predatory. In a number of cases the young are predatory but the adults are not.

- Predation in insects is common. The adults and young of groups such as mantids, dragonflies and damselflies, lacewings, scorpionflies and alderflies are entirely predacious. Even some moths, a group dominated by plant-feeding species, will bear offspring that are voracious predators.

There are also many non-insect invertebrates that are predators. The main group being the spiders, which are specialist predators. Spiders exploit an immense diversity of habitats and have developed some of the most sophisticated trapping and hunting techniques of any animal group. Other predatory invertebrates include scorpions, centipedes, predatory mites, predatory snails and slugs, nematodes and planarian worms.

Invertebrate predators usually feed on other invertebrates, however some feed on fish, frogs, small mammals and birds, in the case of certain large spiders and centipedes. Generally, predators kill to feed themselves but in some cases they hunt for their offspring. Wasps and ants primarily feed on plant liquids as adults but their offspring require solid foods. The adults capture prey and deliver it to their young.

Invertebrate predators usually eat their prey while it is still alive. They do this either with biting and chewing mouthparts or by piercing the body and sucking out the victim's internal fluids. The latter may also involve injecting a cocktail of salivary secretions that act to paralyse the prey and liquefy its tissue.

The methods by which invertebrate predators catch prey are incredibly diverse and in some cases very complex.

Invertebrate predators are either passive or active:

- Passive predators tend to sit and wait for prey to come close. Mantids wait poised for long periods of time for prey to come within the grasp of their powerful forelegs.
- Active predators are those that search or hunt for their prey. Dragonflies and Robber flies use speed and agility to pluck prey from the air, while some spiders construct casting nets that are thrown over prey as they fly past.

Some active predators specialise in capturing and eating other predators. It will vibrate the web of the intended prey in an attempt to imitate a victim caught in it. The web-building spider comes to investigate what they believe is food, to find they themselves are on the menu.

Parasite

A parasite is an organism that lives at the expense of another organism - the host.

In general, parasites share the following features:

- Parasites are usually smaller than their host.
- Parasites use both invertebrate and vertebrate hosts.
- Adult parasites may live on the host), in the host or feed on a host occasionally.
- Parasites generally do not kill the host but may harm the host indirectly by spreading pathogens. This may affect the host's behaviour, metabolism or its reproductive activity.

- Many parasites have hooks, claws or suckers to attach to their host.
- Generally parasites have either a sucker (e.g. leeches) or piercing and sucking type mouthparts (e for feeding.
- Both adults and young can be parasitic. In some cases the young are parasites but the adult is not.

Parasitoid

A parasitoid is an organism that has young that develop on or within another organism (the host), eventually killing it. Parasitoids have characteristics of both predators and parasites.

In general, parasitoids share the following features:

- Parasitoids are usually smaller than their selected host.
- Parasitoids are very selective and only attack a particular life stage of one or several closely related species.
- Adult parasitoids are generally either nectar feeders or predators.
- Only female parasitoids are involved in finding and using the host.
- Generally the host does not die until the young are fully grown and ready to become adults.
- Parasitoids can sometimes prevent larval hosts (e.g. caterpillars) from developing, until the parasitoid is fully developed.

Parasitoids include species of wasps, flies, beetles and worms.

Parasitoids target other invertebrates, although a few are known to infect snake eggs . Parasitoids usually target certain groups, for example spider wasps infect spiders and cuckoo wasps infect other wasps. Some even specialise in targeting hosts that are already infected.

Parasitoids infect their hosts using one of the following three methods:

1. Eggs are laid in, on or near host eggs and young.
2. Eggs are laid on plant species visited by the host.
3. Eggs are laid in, on or near adult hosts, which have been stung and paralysed.

Parasitoid larvae hatch and grow by feeding on the bodily fluids or the internal organs of the host (usually the non-vital parts first). The host dies when the fluid has been sucked dry or its internal organs cease to function.

Natural Enemies of Tea Pests

Various predators, parasitoids and pathogens were found to be effective in controlling tea pests. Tea scientists have earmarked few predators/parasitoids/ pathogens as natural enemies of tea pests which are listed below :

Insect/Mite/Disease/Nematode	Predator/Parasitoids/Pathogen
Black Scale	*Chilocorus circumdatus*
Red Spider Mite	*Stethorus gilvifrons*
	Chrysoperla carnea
	Scolothrips sexmaculatus
	Paecilomyces fumosoroseus
	Verticillium *lecanii*

Contd...

Table Contd...

Insect/Mite/Disease/Nematode	Predator/Parasitoids/Pathogen
Scarlet Mite	*Pronematus* sp. *Exothoris caudate* *Cunaxa* sp. *Agisternus* sp. *Paecilomyces fumosoroseus* Verticillium lecanii
Pink Mite	*Amblyseius herbicolus* *Paecilomyces fumosoroseus* Verticillium lecanii
Tea Mosquito Bug	*Melamphus* sp. *Hexanemus* sp. *Oxyopes javanus* *Beauveria bassinana*
Looper Caterpillar	*Sarcophaga* sp. *Apantales* sp. Argyrophylax sp. *Cotesia* sp. Bacillus thuringiensis
Red Slug Caterpillar	*Exorista* sp. *Argyrophylax* sp. Cotesia sp. *Bacillus thuringiensis*
Eelworm	*Pasteuria neutrans*
Leaf rollers	*Aphanogamus monitae* *Ascogaster* sp.
Flush worm	*Sympiesis dolicogaster*
Green fly	*Dyrid cephalosporium* *Beauveria bassinana*
Bunch Caterpillar	*Bacillus thuringiensis*
Red rot	*Trichoderma harzianum* *Ganoderma pseudoferreum*
Black rot	*Trichoderma viride*

Weed Extract

There are many weeds, the extracts of which can help in controlling tea pests. Few of them are described below :

Name of weeds	Common name
Ageratum conyzoides	Ilame
Polygonum runcinatum	Ratnaulo
Artemisia vulgaris	Titepati
Urtica dioecia	Shisnoo
Eupatorium glandulosum	Ban mara
Equisetum arvense	Kurkure jhar

Note : Soxhlet Apparatus is used for extraction. Petroleum ether, hexane/ Methanol used as solvent.

Other Plant Extracts

Nicotine ($C_{10}H_{14}N_2$)

Pure form of Nicotine is 1-3-(1-methyl – 2- pyrolidyl) pyridine or 1-methyl – 2,3(pyridyl) pyrrolidine. It is derived from *Nicotiana tabacum* and *Nicotiana rustica*.

Pyrethrum ($C_{23} H_{30} O_3$)

Derived from the plants *Crysanthemum cinerariaefolium* and *Crysanthemum coccineum.*

Rotenone ($C_{23}H_{22}O_6$)

Derived from *Derris eliptica* and *Derris malaccensis.*

Anabasine (Neo nicotine)

Derived from *Anabis aphylla* and *Nicotiana glauca*.

Nornicotine ($C_9H_{12}N_2$)

Derived from *Nicotiana sylvestris* and *Daboisia hopwoodi*. Chemically it is 2-2(2-pyridyl pyrrolidine).

Sabadilla

Derived from *Sabadilla officinale* and *Schoenocaulon officiane*. It is mixture of alkaloids veratridine ($C_{36}H_{51}NO_{11}$) and Ceradine ($C_{32}N_{49}NO_9$).

Ryania

Derived from *Ryania speciosa*.

Hellebore

Derived from *Veratrum album* and *Veratrum viride.*

Quassia

Derived from *Quassia amara.*

Picrasmin

Derived from *Picrasma excels.*

Scabrin

Heliopsis scabra and *Heliopsis longipis.*

Nirgundi

Derived from *Vitex nigundo.*

Photographs of Few Weeds Used for Bio-pesticides

Ageratum conyzoides

Artemisia vulgaris

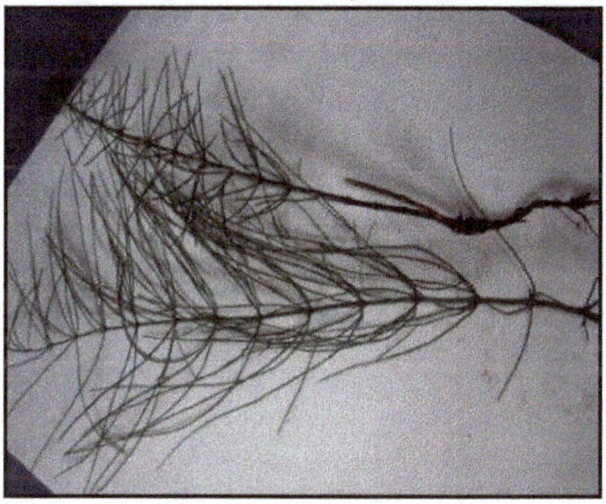

Equisetum arvense

Urtica dioecia

Polygonum runcinatum

Neem extract also is greatly used as insect repellant and antifeedant:

Azadirachta indica (Neem)

Mode of Action : About 150 limonoids have been isolated from Neem plants.Limonoids are a group of homogenous tetranor triterpenoids. More than 30 different compounds viz Betasitosterol, fatty acids,flavonoids are available in Neem. Most important one is azadirachtin. Among these, nine compounds are known to impede growth in insects. Azadirachtin, meliantriol and salanin are feeding deterrants. More than 200 species of insects are found to be affected by Neem deriavatives. Neem acts as antifeedant, oviposition deterrent, insect growth regulator, sterilant and biopesticide.In tea it has fair control (60-70 %) against thrips & jassids and moderate control (50 – 60 %) against mites , tea mosquito bug, looper & bunch caterpillars. Larvicidal activity of Neem is unique. Active ingredients of Neem pesticides are structurally similar to insect hormones called "Ecdysomes". These "Ecdysomes" in insects regulate the process of metamorphosis as the insects pass from larva to pupa to adult. Neem active ingredients soon after their entry into insects' body act as "Ecdysome Blockers", which block the production and release of vital hormone required for reproductive activity. It disrupts mating and sexual communication (release of pheromones) in insects. It deters female from laying eggs and poison eggs (ovicidal activity). It sterilizes adults. It disrupts or prevents moulting thus larva do not develop into pupa or adult. Even if adults are formed, they become weak and abnormal.

Biological Nutrient Management

There should be a combination of bacteria or fungus based fertilizers, which are commonly termed as bio-fertilizers *viz. Azotobacter, Azospirillum,* Phosphorus Solubilizing Bacteria (PSB) or Phosphorus Solubilizing Microorganisms (PSM) like *Bacillus megatherium,* Vescicular Arbuscular Mycorrhizae (VAM) Potash Bacteria *(Frateuria aurentia)* and organic manures like farm yard manure, vermicompost, dried cow dung etc.

Bio Fertilizers

Disadvantage of Powder Formulation of Bio Fertilizers/ Bio Pesticides

1. Shelf life is short.
2. Can not withstand higher temperature
3. Chances of contamination is more
4. Suitable carrier is not available
5. Carriers quality varies
6. Bulk sterilization problem in terms of economy and facilities
7. Mositure absorbing capacity is less.
8. Application cost is more

9. Even under less contamination in packets the enzymatic activity of bacteria is disturbed
10. Packing is a problem
11. Transportation cost is high

Advantages of Liquid Bio Fertilizers / Bio Pesticides

1. Shelf life is more, may be up to 2 years.
2. Contamination probability is less.
3. No effect of high temperature
4. Effective up to 45 degree Celsius.
5. High populations of bacteria/fungus can be maintained ..more than 10^9 cells /ml upto 12-24 months
6. Easy identification by typical fermented smell
7. Dosage is 10 times less than carrier based powder bio fertilizers
8. Very high enzymatic activity since contamination is NIL

Physical Features of Liquid *Azospirillum*

1. The colour of the liquid is either blue or whitish.
2. If the fermentation is not adequate, bad smell is emitted.
3. Colour should be gummy yellow .
4. If the pH is acidic, it becomes difficult for the *Azospirillum* bacteria to survive, hence pH gives an indication of proper medium.

Role of Liquid Azospirillum Under Field Conditions

1. Stimulates growth and imparts green colour which is a characteristics of a healthy plant
2. Helps in utilization of Potassium, phosphorus and other nutrients
3. Increases protein percentage

Production of Growth Hormones

Azospirillum cultures synthesize considerable amount of biologically active substances like vitamins , nicotinic acid, indole acidic acid and gibberellins. All these hormones / chemicals help the plants in better germination , early emergence , better root development

Physical Features of Liquid Phosphorus Solubilizing Micro-organism (PSM) or Phosphorus Solubilizing Bacteria (PSB)

1. Solution turns turbid
2. Colour is generally creamy
3. Smell is pleasant.
4. pH 4.5 or buffer but buffer of 6.5-&.0 ensures more than 2 years shelf-life

Effect of Liquid Phosphorus Solubilizing Micro Organism (PSM) or Phosphorus Solubilizing Bacteria (PSB) Under Field Conditions

1. Facilitates root development
2. PSM synthesizes different kinds of organic acids.
3. Increases the compatibility of other beneficial microbes with the plants
4. Stimulates formation of fats , convertible starches and healthy seeds
5. Induces rapid cell development in the plants and consequently increases the resistance towards diseases
6. Enhances accumulation Mn, Mg, Fe., B., Zn, Cu in the soil.

VAM(Vesicular Arbuscular Mycorrhiza)

1. Obligate symbionts- needs host for multiplication
2. It is difficult to grow VAM in any synthetic media in laboratory
3. Application is difficult.
4. Efficacy varies with variations in soil chemistry.
5. It competes with indigenous VAM in the soil.
6. Slow growth rate of mycorrhizal fungi and slow development of mycorrhizal association

Physical Features of Liquid Potash Mobilizing Bacteria

1. Solution is generally brown in colour
2. Smells like Hydrogen Sulphide or chocolate
3. pH 4.5 but buffer 6.5-7.0 ensures more than two years' shelf life

Role of Liquid Bio-fertilizers in Various Crops

1. Increases crop yield
2. Reduces need of chemical N,P & K fertilizers
3. Benefits to the next crop too, due to its residual effect
4. Enhances soil health and soil fertility
5. Returns are economically viable
6. Besides N.P. & K nutrition, growth hormones secreted increases the crop productivity. Disease and pest occurrence reduced

Role of Biofertilizers in Tea

Curtails the requirement of inorganic fertilizers.

Increases availability of nutrients.

Gives 20 – 50 kg of Nitrogen and 30 – 50 kg of Phosphate in a year per hectare.

Liberates growth promoting hormones vitamins; helps to improve soil fertility.

Suppresses incidence of pathogens; secretes antibiotics to control diseases.

Increases tea yield by 30 – 50 per cent

Mode of Action /Utility

Biofertilizers : *Azotobacter* and *Azospirillum* biofertilizers in non-legumes can replace 25 – 30 % of chemical Nitrogen when used in combination with chemical fertilizers. Phosphorus solubilizing bio fertilizers have been found to be solubilizing 10 – 15 kg soil Phosphorus per hectare per season.. Like *Bacillus megatherium, Pseudomonas* also acts as Phosphate solubilizers. VAM (e.g. *Glomus* spp.) acts as Phosphate absorbers. Mycorrhizal roots lack root hairs. The fungal sheath together with the hyphae extending to the soil absorb nutrients. The Hartig net acts as a liaison tissue between the fungal sheath and the host cells. Mycorrhizal habit increases the surface. Area of the root system and affords better intake of nutrients. It also prevents the root from exogenous pathogenic infections by creating mechanical barrier. PSB secrete many metabolites *viz* CO_2, H_2S,H_2SO_4,HNO_3 and organic acid which are responsible for Phosphorus solubilizing . Apart from fixing atmospheric Nitrogen, *Azotobacter* and *Azospirillum* secrete auxins, vitamins and growth promoting substances and antifungal antibiotics. *Azospirillum* maintains a much more intimate association with host roots compared to *Azotobacter. Azospirillum* is predominant in soil having pH 4.5 to 7.5, while *Azotobacter* thrives well in a pH range of 5.8 to 7.8.While *Azotobacter* requires mediocre soil moisture for survival, *Azospirillum* requires higher moisture content.

Organic Manures /Humic acids / Fulvic acids : To compensate for inorganic fertilizers, use of organic manure is also highly recommended alongwith biofertilizers. Nowadays good quality organic manures are available in the market, which can improve productivity of tea and also improve microbial populations in the soil , thereby enriching its fertility status. Readily available humic acids are also in use in many tea gardens. Humic acids form the bulk of the humus complex and regarded as polymers of aromatic compounds. Humic substances help in better root growth and enhances uptake of minerals. Mobilization of nutrients mainly NPK from the soil to root system is increased in presence of humus substances. Humic acids can reduce phosphorus fixation in soil. Humates have a similar kind of role like EDTA (ethylene-di-amino tetra acetic acid) as far as chelating effect on trace element is concerned. It increases cytochrome oxidase activity in root system. It also increases the activity of glutamic acid transaminase and phosphorylase enzymes. Fulvic acid is the alkali as well as acid soluble part of soil organic matter and contains carbohydrate and proteins.

Merits of Bio-organic Culture

It does not pollute the soil and ground water with chemical residues

It increases the biological diversity of plants and animal

It prevents leaching of minerals in the soil

It utilizes natural equilibrium for crop protection

Do Not Jump to Bio-organic Overnight

There can be a considerable reduction in productivity , if suddenly a tea garden switches over to bio organic form from a conventional form. There can be two distinct methods for such conversion :

(i) Transform to bio organic through the pathways of Integrated Pest Management (IPM) and Integrated Nutrient Management (INM).

(ii) Convert one third of the tea garden into bio organic and then proceed in phasic manner.

Spray of Bio-control Agents

- Spray while there is sufficient moisture in the atmosphere & soil.
- Use a sticker for better result.
- Spray while there is minimal sunshine.
- Wash the spray machines and containers before spray.
- There should not be any spray or broadcast of fungicide or bactericide before & after seven days' of spray.

Post Pruning

To control *Poria* attack, immediately after, pruning, apply *Trichoderma* paste on the bigger cut marks and spray *Trichoderma* @ 1 ml /5 g per litre on the bush frame.

Nutrient Management

In a bio tea garden use of inorganic fertilizers is not allowed. Hence, organic manures and bio fertilizer combination should be used.

Inputs in Planting Pits

At the time of new planting / infilling following ingredients may be used :

Cow dung – 1kg + SSP 30 g + Oil Cakes – 30 g per pit .

Inputs in New Plantation after Planting upto one year

Application of cow dung lassi at monthly interval on the soil of the collar region can be quite useful during droughty periods.

Humik acid liquid can be applied on the soil of the collar region at least twice a year @ 2500 ml per hectare.

Inputs After Pruning

All the pruning litters to be buried inside the soil by digging trench in the interspaces. Add Cowdung @ 1000 g per plant.

General Manuring

Since inorganic NPK will not be allowed in a bio tea garden, organic manures and bio fertilizers should be used extensively . Following proposition can be thought of :

Placement of cow dung @ 1 - 2 kg per plant, by digging a small pit in between two bushes (or by broadcast if availability of workers is a problem.

During rainy season, during rain free days, broadcast of bio fertilizers in the following proportions can be used:

Azotobacter @ 2.5 kg + Phosphorus solubilizing bacteria @ 5 kg + Vescicular arbuscular mycorrhizae @ 5 kg per hectare can be applied.

Green Cropping

In the new plantation area, green crops and cover crops like *Crotalaria anagyroides*,, *Tripsacum laxum*(Guatemala grass) , *Mimosa invisa* etc should be planted to increase organic matter content of the soil.

Nutritional Values of Various Organic Matters

	N %	P %	K %
Farm Yard Manure	0.5	0.2	0.5
Tea pruning	2.5	0.5	1.5
Tea waste	3.6	0.6	2.3
Oil Cakes			
Castor	5.0	2.0	1.0
Ground Nut	7.4	1.5	2.4
Neem	5.2	1.0	1.4
Linseed	4.7	1.7	1.3
Coconut	3.4	1.5	2.0
Sesamum	5.5	2.1	1.3
Loppings			
Tripsacum	1.6	0.6	1.7
Napier	0.8	0.4	1.3
Mimosa invisa	2.6	0.6	1.0
Thatch grass	0.5	0.1	0.5
Mulching			
Tea prunings	2.5	0.5	1.5
Tea waste	3.6	0.6	2.3
Paddy straw	0.5	0.07	0.6
Thatch grass	0.5	0.1	0.5
Guatemala grass	1.6	0.4	1.3

Avenue Plantation

On the boundary of each section, Neem, Karanj (*Pongamia*) and madar (*Erythrina indica*) type of plants should be planted at 10 m spacing. In between these trees , small herbs like *Leucaenea leucocephala* (Su babool) , *Sesbania* (dhaincha) should be planted at 1 m distance. These will add sufficient quantity of bio mass in the soil.

Shade Plantation

Shade trees contribute about 15 tonnes of leaf litter per year . Hence leguminous shade trees should be planted in all the sections at recommended spacing. Shade tree loppings can provide nutrients as furnished below:

Nutrient	kg/ha
Nitrogen	63 to 126
P_2O_5	18 – 36
K_2O	22-44
Ca	32-64
MgO	16-32

Rock Phosphate/Wood Ash

Rock phosphate is allowed in a bio tea garden. P_2O_5 @ 50 kg per hectare per year in the form of Rock Phosphate can be applied. Wood ash is a good source of Potassium in organic form. Apart from using Potash bacteria - *Frateuria aurentia* @ 2.5 kg per hectare, wood ash can be used to replenish Potassium in the soil. Wood ash can be obtained by burning shade tree loppings etc. Dose can be ascertained by determining the Potassium content of the wood ash in a recognized laboratory.

Weed Control

No herbicide can be used in a Bio Organic garden. Only manual weed control is allowed.

During winter, cheeling (cheel hoeing) operation should be undertaken in all the sections . In case of heavy weed infestation, chop hoeing can be done. In rainy seasons weeds should be controlled by sickling. Weed control is easy in skiffed sections where canopy itself prevents entry of sunlight to the soil. In Young tea sections , mulching should be done to control weeds.

Extensive research is going on to evolve biological control of weeds by using fungal preparations. When such bio-control agents will be available, weed control biologically can be easier.

Foliar Spray

Bio Foliar spray of sea weed extract @ 500 ml per hectare can be safely used to provide nutrients to the plants.

Foliar spray of Humic Acid (or a combination of Humic acid and fulvic acid) @ 500 ml per hectare can be quite useful.

Bio-dynamic Preparation

Now-a-days bio-dynamic preparations are gaining ground in the Bio-organic tea gardens. Few examples are given below :

Bio-dynamics - 500 (Cow Horn Manure)

Cow dung is stuffed inside the cow horns and buried in a pit. The depth of the pit should be 45 cm. The pit is covered with good quality compost manure.

The distance between two cow horns should be at least 5 cm. The open end of the cow horns should be at the bottom so that no water accumulation is possible. The pit should be mulched by paddy straw and a thatch shade should be constructed on top of the buried cow horns. After about 4 months the dung inside the cow horn turns brown it emits a sweet smell. It should be then taken out of the soil and stored in a cool place and must be kept in glass jars surrounded by damp coir pith.

Dose : 100 gms of this brown cow dung should be sprayed on the tea foliage dissolving in 60 litres of water. Before sprayed it should be stirred vigorously

Time of spray : Post monsoon

Bio-dynamics 501 (Cow horn Silica):

In this method, instead of cow dung, finely ground quartz crystals of silica are inserted inside the cow horn. This preparation is done during hot summer months. Other methods are same as Biodynamics 500. The color of mature horn silica is grey or yellow. It is stored in a glass jar on a sunny window site.It can be stored for 2-3 years.

Dose : 2.5 gms per litre of water. Before sprayed it should be stirred vigorously

Time of spray : July to October.

Bio-dynamics 502: is made from the flowers of yarrow (*Achillia millefolium*). The dry flowers of yarrow are put in stag bladder until the bladder is packed, then stitch the slit with cotton thread. Store in a closed basket to keep away rodents or pests. Hang up in March to get cosmic influences till September. Burry it in September in a mud made flower pot surrounded with fertile soil. After 4-5 month it may be taken from the soil in February or March. It must be stored in cool, dark and moist place.

Bio-dynamics 503: is made from the flowers of German Chamomile (*Matricaria chamomilla*). Cow or Bull small intestine is needed for this preparation. The intestine at first cut into 15cm bits and tie the cut bits at one end by cotton thread. Then put a funnel to the open end and fill with dry flowers of chamomile. The packing should not be to hard or loose. Packed intestine just looks like sausages. Stack the sausages into a bundle which could be placed in a mud pot which must be surrounded with fertile soil or old compost. Bury it in October and let it remain in the soil till Feb March.

Bio-dynamics 504: is made from the leaves of Stinging nettle (*Urtica parviflora*). Fill up the dry leaves in mud pots. Press well into the containers; ensure that the lid is on. Place the pot under the influence of Mars. Place it into soil in early September and lift the preparation after one year.

Bio-dynamics 505: is made from the bark of the Oak (*Quercus glauca*) tree.The bark must be collected from minimum 25 years old oak tree. Finely crush oak bark put in cow skull. Block the opening with a well shaped bone pieces or stone. Place the skull in a watery environment. The preparation is placed in September and lifted in March.

Bio-dynamics 506: is made from the flowers of Dandelion (*Taxaricum officinale*). Place the dry flower in a mesentery of a cow. Ensure that the extra fat is cut off. Do not wash the mesentery and wrap it like a parcel and tie with a jute thread. Place the parcel in a good mixture of soil and compost into a pot. Place the preparation in September and Lift it in March. This preparation is related mainly with salicylic acid. It helps to draw energy from the planet.

Bio-dynamics 507: is made from the flowers of Valerian (*Valeriana officinalis*). It is the liquid preparation among all the Biodynamic preparation. The clipped flowers put into a mortar and grind to form a paste. This paste is added to water in the ratio of **1:4** in a bottle. Store in a cool place. It contains **5** per cent solution.

Biodynamics 502 to 507 are used in compost making

Insect Growth Regulators

These chemicals interfere with the secretion of endocrine glands and thus interfere with chitin synthesis, an essential component of cuticle *e.g.* Diflubenzuron.

Sex Pheromones

Now a days, many types of sex pheromones are being experimented at many tea research centres to ascertain its efficacy in controlling tea pests.

Different Bio-control Agents (Entomo Pathogens and Phyto Pathogens)

Paecilomyces fumosoroseus : Controls mites including red spider mites.

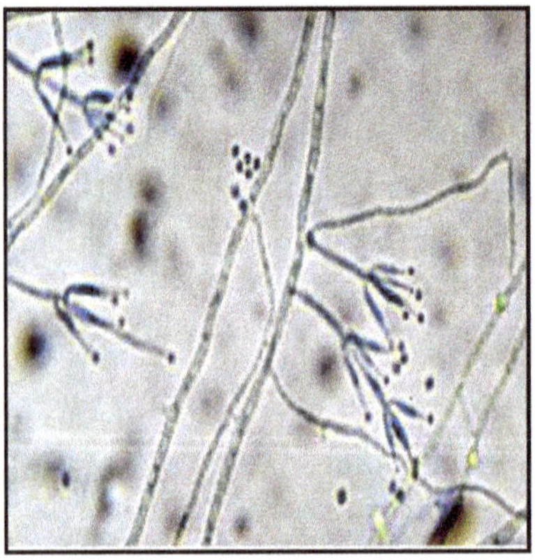

Characteristic features : Conidia is long ovoid in shape. Pink grey , pin tan, tan, grey, reverse pale to yellow in colour. Conidiophore is smooth and colourless; conidial chain long and conidial head are diffuse.

Mode of Action : The spores of *Paecilomyces fumosoroseus* comes in contact with the pests, germinates , multiplies and penetrate into the haemocoel. It takes nutrient from the haemolymph and colonize the whole visceral mass.

Paecilomyces lilacinus : Controls nematodes.

Characteristic features : Conidia is short fusoid, lemon shaped. Purple grey , tan or grey in colour. Conidiophore often roughened and slightly coloured; conidial heads compact with short conidial chains.

Mode of Action : *Paecilomyces lilacinus* produces hyphae, which infects nematodes like *Meloidogyne incognita, Rotylenchulus reniformis, Helicotylenchus incisus, Pratylenchus coffeae* etc. *Paecilomyces* is a facultative parasite of root knot nematode eggs. It disrupts the lip and chitin layer of egg shell.

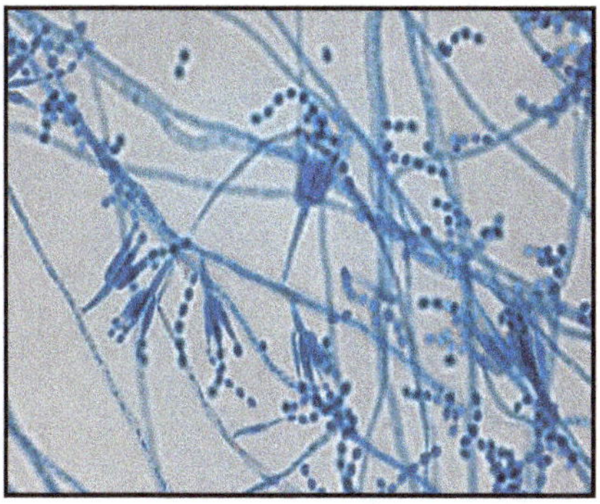

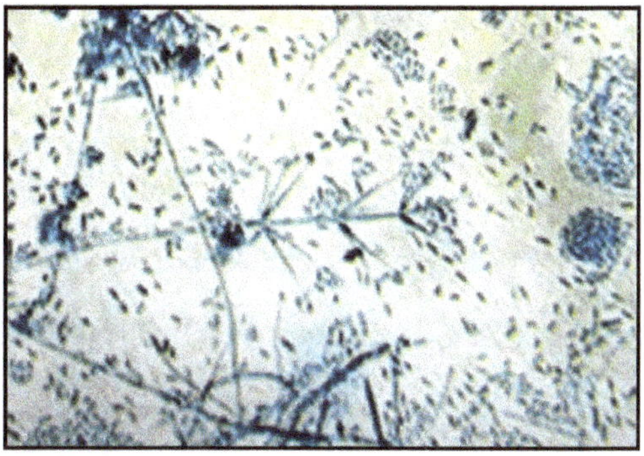

Verticillium lecanii :

Controls mites, aphids,thrips and scale insects.

Characteristic features : Conidia cylindrical with rounded ends. Conidiogenus cells narrow, awl-like, tapering towards tip.

Mode of Action : *Verticillium* enters the body of the insects through insect cuticle and sporulates inside to produce a toxin called bassianolide.

Beauveria bassiana :

Controls sucking pests like flush worms, tea mosquito bugs etc. It can also control looper caterpillars to some extent.

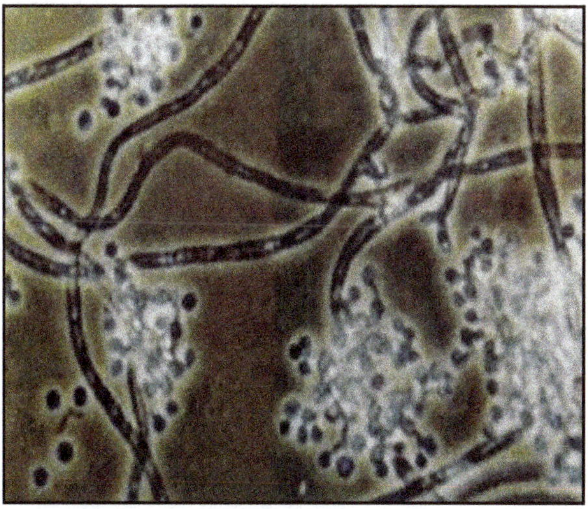

Characteristic features : Conidia globose or sub globose in shape. Pink-grey, pink-tan, grey in colour. Conidiophore smooth and colourless. Conidial chains often long and conidial heads diffuse.

Mode of Action : The spore of *Beauveria bassiana* germinates while it comes into contact with pests. It penetrates through cuticle and multiply inside. It produces a toxic metabolite called Beauvericin. This product inhibits synthesis of metabolic enzymes and leads to paralysis of physiological functions.

Metarhizium anisopliae: Control soil borne insects like termite, cockchafer etc.It can also control pupae of looper caterpillars.

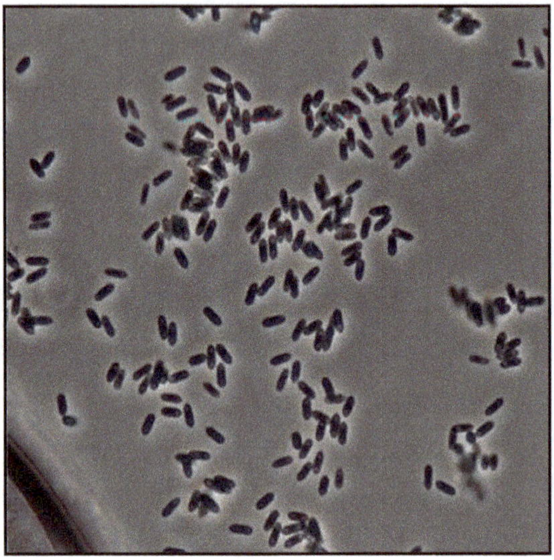

Characteristic features : Conidia cylindrical with middle constricted, green , yellow-green or brown coloured. More often affecting hosts in soil rather than aerial parts.

Mode of Action : *Metarhizium* produces a number of secondary metabolites acting as mycotoxins like Destruxin E. These toxins cause titanic paralysis and rapid development of mycotoxins causes death of termites.

Trichoderma viride :Antifungal/ cures root diseases ; revives plants suffering under waterlogged condition.

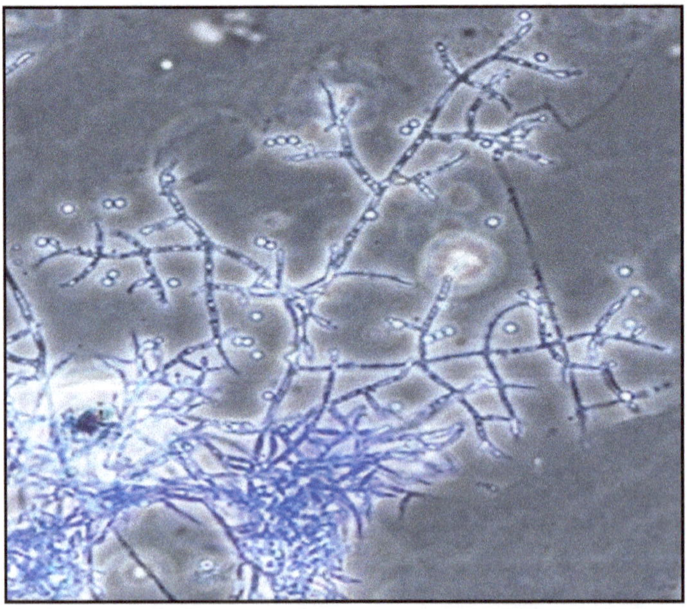

Mode of Action : The conidia of *Trichoderma viride* come into contact with target host. It germinates and the germ tube coils the hyphae of pathogenic fungi , degrades the cell wall and feeds on the nutrients causing mycoparasitism. It also produce toxic metabolites.

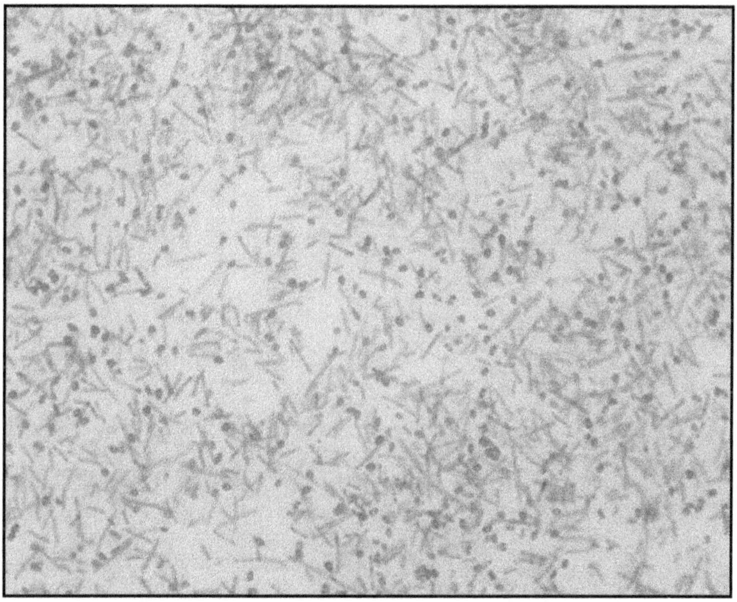

Pseudomonas fluorescence: **Against diseases.**

Mode of Action : *Pseudomonas fluorescens* controls diseases by producing antibiotics and also obstruct supply of food to the pathogenic fungi.

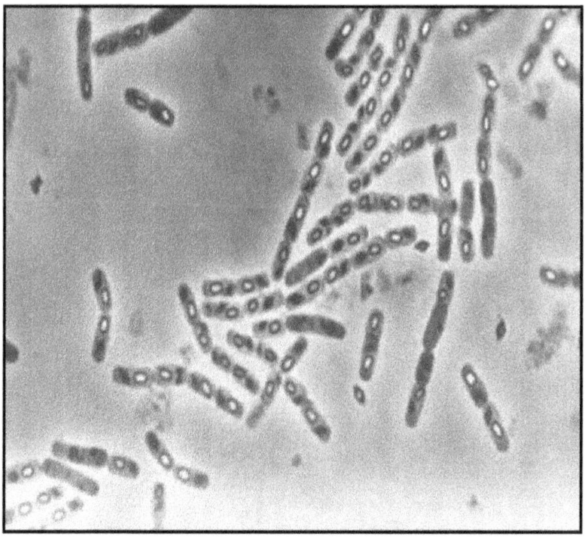

Bacillus thuringiensis: Used against looper and other caterpillars.

Mode of Action : B. thuringiensis produce a class of insecticidal small molecules called beta-exotoxin, the common name for which is thuringiensin. Beta-exotoxin and the other Bacillus toxins contribute to the insecticidal toxicity of the bacterium to lepidopteran.

BIO-CONTROL AGENTS AVAILABLE IN THE MARKET

	Trade Name	Company	Dose	
Azadirachtin 0.03 %	Pestineem	S.B.Industries	1.5 ml/litre	Antifeedant & Repellant
	Multineem	Karnataka Agrochemicals	2 ml/litre	Antifeedant & Repellant
	Pestoneem	Amit Biotech	2 ml/litre	Antifeedant & Repellant
Azadirachtin 0.15 %	Neemgold	Foliage Crop Solutions Pvt. Ltd.	2 ml/litre	Antifeedant & Repellant
	Nimbecidine	Indofil Industries Ltd.	2 ml/litre	Antifeedant & Repellant
	Phytoneem	WBPPDCL	2 ml/litre	Antifeedant & Repellant
	Achook	Godrej Agrovet Ltd.	2 ml/litre	Antifeedant & Repellant
	Nimactin	Biostadt India Ltd.	1.5 ml/litre	Antifeedant & Repellant
Azadirachtin 3 %	Econeem	Margo Biocontrols Pvt. Ltd.	2.5 ml/litre	Antifeedant & Repellant
	Azacel – 3000	Excel Crop Care Ltd.	3 ml/litre	Antifeedant & Repellant
	Ultineem 3000	Super Agro Ltd.	3 ml/litre	Antifeedant & Repellant
Azdirachtin 0.5 %	Ultineem 5000	Super Agro Ltd.	2 ml/litre	Antifeedant & Repellant
	Ecotin	Margo Biocontrols Pvt. Ltd.	2 ml/litre	Antifeedant & Repellant
Azadirachtin 1%	Phytoneem 10000	WBPPDCL	1 ml/litre	Antifeedant & Repellant
	Econeem Plus	Margo Biocontrols Pvt. Ltd.	1 ml/litre	Antifeedant & Repellant
	Ultineem 10000	Super Agro Ltd.	1.25 ml/litre	Antifeedant & Repellant
Azdirachtin 5 %	Neemazal – F	E.I.D.Parry	0.3 ml/litre	Antifeedant & Repellant
	Neem Fol	Gassin Piere	0.3 ml/litre	Antifeedant & Repellant
	Nimbecidine Plus	Indofil Industries Ltd.	0.3 ml/litre	Antifeedant & Repellant
Bacillus thuringiensis	HIL-BTK	Hindustan Insecticides Ltd.	2-3 gms/litre	Caterpillars
	Panther BT	Leibig's Agro Chem Pvt. Ltd.	2-3 gms/litre	Caterpillars
	Dipole	Amit Biotech	2- 3 gms/litre	Caterpillars
	Halt	Biostadt India Ltd.	2- 3 gms/litre	Caterpillars

Contd....

Table Contd....

	Trade Name	Company	Dose	
	Dipel- BL	Sumito Chemicals India Pvt Ltd.	2- 3 g/litre	Caterpillars
Beauveria bassiana	Bio Power	Indofil Industries Ltd.	3-6 g/litre	Sucking pests
	Larvocel	Excel Crop Care Ltd.	5 g/litre	Sucking pests
	Wah	Amit Biotech	6 g/litre	Sucking pests
Metarhizium anisopliae	Terokil	Amit Biotech	10 ml/litre	Termites/ Cockchafer grub
	Biotercel	Excel Crop Care Ltd.	20 g/litre	Termites/ Cockchafer grub
Paecilomyces fumosoroseus	Priority	Indofil Industries Ltd.	5 g/litre	Mites
Pseudomonas fluorescence	Sudocel	Excel Crop Care Ltd.	5 g/litre	Mites and other sucking pests
	Rakshak	Amit Biotech	5 g/litre	Mites and other sucking pests
	Nima – II	Krishi Rasayan	5 g/litre	Mites and other sucking pests
Sophora flavescence	Drone	Chemtura Chemicals India Pvt. Ltd.	1 ml/litre	Caterpillars
Trichoderma viride	Tricho XP	Excel Crop Care Ltd.	2.5 g/litre or 2.5 -5 kg per hectare	Antifungal
	Nisarga	Karnataka Agrochemicals	7.5 kg – 10 kg per hecatre	Antifungal
	Guard	Amit Biotech	2.5 g/litre or 5 kg per hectare	Antifungal
	Biocure F	Indofil Industries Ltd.	3 kg per hectare with 500 kg FYM	Antifungal
	Kesana	Krishi Rasayan	2.5 kg /hectare	Antifungal
	Tricostar	Super Agro Ltd	1 kg per 150 kg FYM	Antifungal
	Nisarga Liquid	Karnataka Agrochemicals	0.5 ml/litre	Antifunga
Verticillium lecanii	Verticel	Excel Crop Care Ltd.	2.5 g/litre	Mites
	Kilmit	Amit Biotech	7 g/litre	Mites
	Biocatch	Indofil Industries Ltd.	5 g/litre	Mites

Calendar of Methodology for a Bio-organic Tea Garden

January

Foliar Spray

One round of Neem and Sulphur combination as a prophylactic spray against mites and sucking pests. Neem @ 1000 ml + Sulphur @ 2.5 kg per hectare.

February

Foliar Spray (Bio-pesticides)

Paecilomyces fumosoroseus @ 250 ml + Neem @ 1000 ml per hectare.

March

Foliar Spray (Nutrients and Growth Promoters)

Sea Weed Extract @ 500 ml + Humik Acid @ 500 ml per hectare.

Foliar Spray (Bio-pesticides)

Verticillium lecanii 250 ml + *Paecilomyces fumosoroseus* @ 250 ml per hectare.

April

Foliar Spray (Bio-pesticides)

Beauveria bassiana @ 100 ml + Neem @ 1000 ml per hectare.

May

Foliar Spray (Nutrients and Growth Promoters)

Sea Weed Extract @ 500 ml + Humik acid @ 500 ml per hectare.

Ground Application (Nutrients)

Placement of Cowdung @1.5 tonnes + Vermicasting @ 30 kg *Trichoderma* @ 2.5 kg (light watering/irrigation after application if sufficient moisture is not present in the soil).

June-august (Monsoon)

Need based application.

Foliar Spray (Bio-pesticides)

Beauveria bassiana @ 100 ml + Neem @ 1000 ml per hectare once in each month.

Foliar Spray (Bio-pesticides)

Pseudomonas fluorescence @ 250 ml per hectare

September

Foliar Spray (Nutrients & Growth Promoters)

Sea weed extract@ 500 ml + Humic acid formulation@ 500 ml per hectare.

Ground Application (Nutrients)

Placement of Cowdung @1.5 tonnes + Vermicasting @ 30 kg *Trichoderma* @ 2.5 kg (light watering/irrigation after application if sufficient moisture is not present in the soil).

October

Foliar Spray (Bio-pesticides)

Paecilomyces fumosoroseus @ 250 ml + Neem @ 1000 ml per hectare.

November–December

No application

Note :

(i) In case of looper attack use *Bacillus thuringiensis* at recommended doses as and when required.

(ii) Bio control agents, mainly entomopathogens should be used only if pest is present.

(iii) Biodynamic formulations can be applied as foliar nutrients

Pest Surveillance

"Prevention is better than cure", "A stitch in time saves Nine" and " Nip in the bud"- these three proverbs are very appropriate in tea gardens for developing a well-knit pest control mechanism. Before taking out the guns to eradicate pest problems, it is pertinent to know the kind of pest attack and its intensity through effective feed back system. Like a successful warrior, it is important to get a clear picture about the strength of the enemy and then formulate a full-proof war strategy. In a tea garden, there should be a method to regularly study the intensity of pest attack in each tea section through PEST SURVEILLANCE technique. An Assistant Manager in a tea garden can be given the responsibility to visit each and every section once a week and prepare a Pest Surveillance color chart. A design of the chart is furnished below as ready reference:

RED	Severe infestation, when immediate spray is required.
YELLOW	Considerable infestation but spray of pesticide may be deferred.
BLUE	Moderate infestation, to be decided if pesticide spray is at all required or not.
GREEN	Pesticide spray not required.

PEST SURVEILLANCE

Week :

Section #	Year of Planting	Area (ha)	Red spider mites	Thrips	Jassids	Tea Mosquito Bug	Caterpillars	Cockchafer	Red rust	Other pests

RED	Severe infestation, when immediate spray is required.
YELLOW	Considerable infestation but spray of pesticide may be deferred.
BLUE	Moderate infestation, to be decided if pesticide spray is at all required or not.
GREEN	Pesticide spray not required.

From this chart, it will be easy to prepare pesticide spray schedule. The 'red' marked sections should be sprayed first and then the 'yellow' sections. In case of 'red' marked sections also the priority should be given on the basis of the species of the insect / mite and causal organism of fungus. If the section is infested by looper caterpillars or red slug caterpillars, it should get top priority rather than red spider mites or scarlet mites.

Compost Pit

- Construct a tank of the size : length 3 metres, width 2 metres and height 1.5 metres with pigeon holes.
- Put a plaster of liquid cowdung inside the tank like plastering huts.
- Spread a polythene sheet on the floor.
- Spread green leaves, agro waste etc about 125 – 150 kg to the height of 6 inches.
- Mix about 10 kg dung in 100 litres water and spray on the greeneries.
- Spread about 125 – 150 kg of cattle urinated soil on top of it.
- This process to be continued till the tank is filled.
- Cover the tank with polythene sheet.
- After 10-15 days fill the gap in the same manner.
- Within 40 – 50 days compost will be ready.

Note: one tank can produce 12 M.T. of organic manure in a year.

Vermiculture

Worms Used

- *Eisenia foetida*
- *Amyanthas diffringens*
- *Eudrilus eugeniae*
- *Perionyx excavatus*
- *Lampito mauritii*
- *Pheretima elongate*
- *Pontoscolex corethrurus*

Photographs of Earthworms

Eudrilus eugeniae

Perionyx excavatus

Eisenia foetida

Amyanthas diffringens

Nutrient Content of Vermicompost

Nutrients	Content (approximately)
N (%)	1.5
NH_4-N (ppm)	85
NO_3-N (ppm)	100
P_2O_5 (%)	0.75 – 1.12
K_2O (%)	0.65 – 1.14
S (%)	0.24-0.52
pH	6.76 – 7.28
C/N Ratio	12
Humic acid	5
CEC(cmol (p+)kg^{-1}	95
Zn (ppm)	55

Rearing of Worms

- Wooden box of 45 cm X 65 cm X 20 cm with drainage holes.
- Rice husks, cereal straw, saw dust are used as breeding material.

- Ferment for one month.
- Mix Guatemala grass / water hyacinth / poultry manure.
- pH should be neutral, temperature 20 – 27 ^{0}C .
- 100 g of worms from breeder's stock required for each 0.1 m ^{2} of surface area.

Composting

- Pit size – 2 m x 1 m x 1 m.
- Bricks and sand at the base.
- Line the pit with old gunny bags.
- Fill the pit with FYM, green weeds and other organic residues.
- A layer of 15 cm. Loamy soil to be sprinkled and keep moist.
- Two third of the worms from rearing box to be harvested and placed on top of the heap.
- 10 kg of vermicasting is produced per kg of worm.

11

SHADE TREES

Shade trees have become the integral part of the tea gardens in the plains because of the following benefits:

(1) It cut off solar insolation which is detrimental to tea plants.

(2) It does not create high amount of darkness, it allows 50 per cent of the sunlight to pass in.

(3) Being leguminous plants (mostly from the family *Fabaceae),* their roots can harbour nitrogen fixing bacteria. This adds nitrogen in the soil.

(4) Since leguminous shade trees are deciduous, it sheds leaves once in a year, fallen leaves adds humus and helps in nutrition.

(5) It creates a micro climate in the phyllospheric region of the tea plants, conserve moisture and helps the plants to continue normal photosynthesis.

(6) Regulated shade reduce the infestation of pests which are positively thermo and phototropic viz. mites, thrips etc.

(7) Shade trees serves as wind/storm barrier.

(8) The roots of shade trees are strong and highly proliferated, thereby it can hold the soil and reduce soil erosion.

(9) There are many shade trees *e.g. Albizzia adoratissima,* wood of which is used for making furniture.

(10) Twigs of shade trees are used as firewood.

(11) Shade trees provide shade which increases the productivity and working energy of the pluckers and other workers during hot summer days.

Characteristic Features of Few Major Shade Trees

Albizzia chinensis **(Osbeck) Merr syn stipulata Boiv**

Family: Fabaceae

Common Name : Sau Sirish

Category : Semi-permanent shade tree

A medium sized tree. Maximum height about 15 metres. Bark colour is light brown.

Leaf : Compound leaf, bipinnate, paripinnate. Length of leaf rachis is about 30 cm with a gland at the base of petiole. Large stipules, about 2 cm long is the characteristic feature of this plant. Pinnae 10-20 pairs, 5-18 cm long. Leaflets are sharp pointed at the apex. Leaflets are sessile, 30-50 pairs , 0.7 to 1.5 cm long and 0.2 to 0.4 cm broad.

Fruit : Length of the pod is 12-15 cm and width is 1.5 to 2 cm. The colour of the pod is straw colour when ripe. In each pod there are 4 to 10 seeds present.

Leafless Period : End January to end March. Hence this shade tree is not very suitable in a droughty area.

Density of shade : Light to medium

Albizzia odoratissima **Benth.**

Family : Fabaceae

Common Name : Kala Sirish

Category : Permanent shade tree

A large deciduous tree with spreading crown. The colour of the bark is grey.

Leaf : Pinnately compound , bipinnate, paripinnate. Length 10 to 25 cm. Pinnae 4 to 8 pairs. Each pinnae 8 to 15 cm long. Leaflets 6 to 16 pairs ,dark green in colour, 1 to 2 cm long and 0.4 to 0.7 cm wide.

Fruit : Length of the pod is 15 to 30 cm and width 3 to 4 cm. The colour of the pod becomes dark brown when ripe. In each pod there are 8 to 12 seeds present.

Leafless period : End December to February. This shade tree is very suitable in the droughty areas. In fact this should be the first choice as a shade tree.

Albizzia odoratissima **plants in the nursery**

Legume (pod) of *Albizzia odoratissima*

***Derris robusta* Benth.**

Family : Fabaceae

Common Name : Derris, Sirkhi.

Category : Permanent shade tree.

A medium sized tree . The height goes upto 15 metres. The colour of the bark is whitish grey.

Leaf : Simple unipinnate, imparipinnate. Length 15 to 20 cm. Leaflets 10-25. Length of each leaflet is about 2 to 3.5 cm and width about 1 to 2 cm. Apex of each leaflet is mucronate.

Fruit : The length of the pod is 2.5 to 5 cm and 0.5 to 0.7 cm wide. The pods become straw coloured when ripe. Number of seeds in each pod is 1 to 4.

Leafless Period : Mid December to February. This shade tree is very suitable in a droughty area.

Sometimes the *Derris robusta* plants serve as the host of looper caterpillars but if proper pest control measure is taken, this species of shade tree can be very useful in a tea garden.

Albizzia lebbek **(L.) Benth.**

Family : Fabaceae

Common Name : Lebbek, Lasrin.

Category : Permanent shade tree.

A medium sized tree going upto the height of 15 metres. The colour of the bark is light grey to dark grey.

Leaf : Pinnately compound , bipinnate, paripinnate. Length 7 to 15 cm. Pinnae 3 to 6 pairs. Each pinnae 10 to 15 cm long. Leaflets pale green in colour, 3 to 8 pairs , 2 to 5 cm long and 1 to 2.5 cm wide.

Fruit : Length of the pod is 15 to 30 cm and width 3 to 4 cm. The pod becomes straw coloured when ripe. In each pod there are 6 to 10 seeds present.

Leafless period : January to April. This shade tree is not very suitable for the droughty area.

A twig of *Albizzia lebekk*

A legume of *Albizzia lebekk*

Acacia lenticularis Buch-Ham.

Family : Fabaceae

Common name : Kanta sirish.

Category : Permanent shade tree.

The tree is about 15 to 20 metres high. The characteristic feature of this shade tree is it has hooked spines. The colour of the bark is grey.

Leaf : Pinnately compound , bipinnate, paripinnate. Length 12 to 20 cm. Pinnae 12 to 16 pairs. Each pinnae 10 to 20 cm long. Leaflets pale green in colour, 12 to 16 pairs , 2 to 2.5 cm long and 0.7 to 1 cm wide.

Fruit : Length of the pod is 15 to 20 cm and width 2.5 cm. The pod becomes dark brown coloured when ripe. In each pod there are 5 to 8 seeds present.

Leafless period : March to April. This shade tree is not very suitable for the droughty area. But this shade tree is very suitable in the area which has cattle trespass problem.

***Adenanthera pavonina* Linn.**

Family : Fabaceae

Common Name : Rakto chandan

Category : Permanent shade tree.

The height of the tree 5 to 10 metres. The colour of the bark is brownish grey.

Leaf : Pinnately compound , bipinnate, paripinnate. Length 20 to 40 cm. Pinnae 3 to 6 pairs. Each pinnae 7 to 10 cm long. Leaflets 6 to 16 pairs , 2 to 3 cm long and 1 to 2 cm wide.

Fruit : Length of the pod is 15 to 20 cm and width 1.5 cm. The pod splits up when ripe. In each pod there are 8 to 16 seeds present. The colour of the seeds is bright scarlet red .

Leafless period : January to March.

Indigofera teysmanni Miq.

Family : Fabaceae

Common Name : Indigofera

Category : Temporary shade tree.

This is a small tree about 4 to 5 metres in height.

Leaf : Pinnately compound, unipinnate, imparipinnate. Length 20 to 30 cm. Leaflets lanceolate , apex and base tapering, 8 to 15 pairs , 5 to 8 cm long and 3 to 4 cm wide.

Fruit : Pod is cylindrical. Length of the pod is 2,5 cm and width 0.3 cm. The pod becomes blackish brown coloured when ripe.

Leafless period : Mid January to Mid March.

Indigofera teysmanii

There are many other shade trees which are useful in tea industry viz *Leucaena leucocephala* (Subabool) etc.

Planting Plan of Tea and Shade Trees

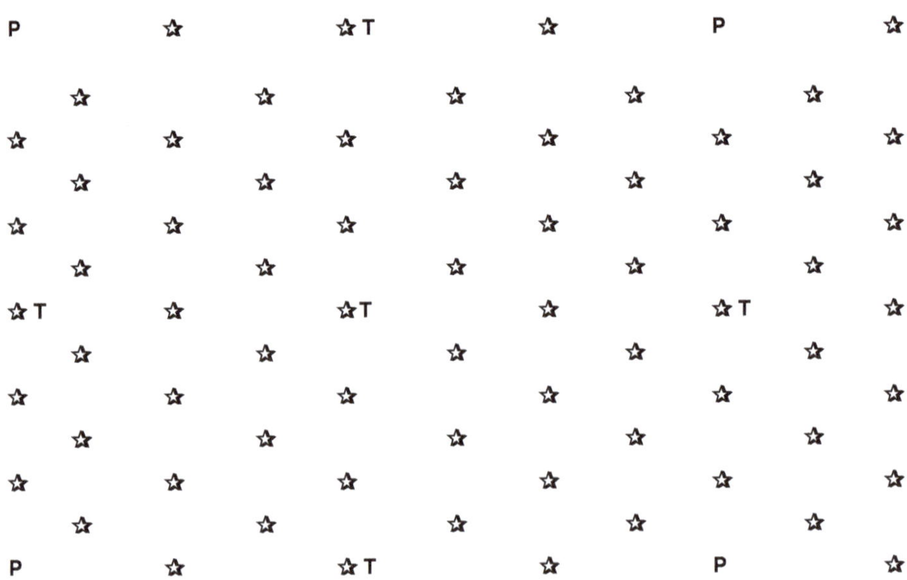

P = Permanent Shade Tree T = Temporary Shade Tree * = Tea plant

Distance between plant to plant (tea) : 60 cm

Distance between row to row (tea) : 105 cm

Spacing of permanent shade trees	: 735 cm X 720 cm
Position of permanent shade tree	: In place of a tea plant.
Position of temporary shade trees	: In between permanent shade trees.

12

PRUNING

Pruning is one of the most important operations in a tea garden. The objectives of young tea pruning and mature tea pruning are different. In case of young tea pruning is done to induce lateral growth for producing more plucking points which in turn helps in increasing productivity. Mature tea pruning is done with the following objectives :

- To stop reproductive growth and encourage vegetative growth.
- To remove knots, unproductive diseased woods and change gnarled appearance of the bush.
- To provide a height which suits the pluckers.

Time of Pruning

Mature tea :	Mid- November to mid-January
Young tea :	End January to early February

(Maximum starch reserve found in February)

How to Decide Pruning Time ?

1. **Movement of photosynthates**: Direction of movement of photosynthates from the maintenance foliage is decided by the growing stage of shoots. When the shoots are growing, the photosynthates move basically in upward direction to the growing shoots. But when the buds grow dormant when there are no growing shoots the direction of the movement of photosynthates is diverted towards the roots.
2. **Starch reserve and pruning time:** Starch reserved decides pruning time for the young and mature tea considering the bidirectional flow commencing October. This may be the earliest time for resting the bushes for prior to pruning another bidirectional movement occurs in May, which may be

utilized for decentering in young tea. It may be noted that presence of good starch reserve in the roots is an important factor to reduce dieback and to have quicker and better recovery of bushes.

Starch Test

Iodine crystals (3 g) and Potassium Iodide (6 g) in one litre of water or weak hospital iodine solution may be painted on the newly cut, pencil thick root. Dark complexion reflects adequate starch reserve. If the colour does not change, it will denote that starch reserve is not adequate.

Young Tea Pruning

This aspect has been discussed in details in the YOUNG TEA MANAGEMENT CHAPTER.

Terminologies Of Pruning

FFP = Frame forming pruning

CP = Collar pruning

RP = Rejuvenation pruning

HRP = Height reduction pruning

MP = Medium pruning (Cut back pruning)

LP = Light pruning (Cut across pruning)

DS = Deep Skiff

MS = Medium Skiff

LS = Light Skiff

LOS = Levelling – off - Skiff

GL = Ground Level

Mature Tea Pruning

Height of Pruning

Collar Prune : 10 cm from Ground level (GL)

Heavy Prune / Rejuvenation Prune : 40 – 45 cm in plains and 15 – 37 cm in hills from GL

Medium Prune : 50-65 cm in plains and 35 – 50 cm in hills from GL.

Height Reduction Pruning : 70 – 75 cm in plains only from GL.

Light prune : 4 – 5 cm above last pruning mark.

Deep Skiff : 10 – 13 cm above last LP mark.

Medium skiff : Just below majority of crow's feet.

Light skiff : At current years tipping level.

Levelling – off – skiff : 5 cm above current year's tipping level.

Pruning Cycles

IN THE PLAINS TEA GARDEN

3-4 years : LP-UP-UP ; LP-DS-UP ; LP-UP-DS ; LP-MS-DS ; LP-UP-DS / MS –UP

IN THE HILLY AREAS

4 – 5 years : LP-UP-DS-LS-UP ; LP-UP-UP-DS / MS –UP

Tipping heights

First plucking after pruning is known as Tipping

CP	: 70-75 cm from the ground level
MP/HRP	: 25-30 cm from the pruning mark
RP	: 25-35cm from the pruning mark
LP	: 20-25cm (avg. 5 leaves) from the pruning mark
DS	: 7-10 cm (avg. 2 leaves) from the pruning mark
MS	: 4-5 cm (avg. 1 leaf) from the pruning mark
LS / LOS	: Janam (scale leaf) level

Pruning Knives

For LP	: 15 cm blade
For DS /MS	: 25 cm blade
For LS / LOS	: 30 cm blade
Cleaning Out	: 7.5 – 10 cm blade

The weight of the LP knife should be not less than 450 g.

Knife Cleaning Out (Kco)

1) All the criss cross branches , thin whippy branches, one of the two branches at the close proximity, hanging branches to be removed.
2) Old pruned marks affected by *Poria hypobrunnea* should be scraped upto the clean wood.
3) A 7.5 cm blade hook type pruning knife should be used for KCO

Benefit of Breathers

After LP / MP/HRP/RP it is better to keep a breather in each pruned bush.

- Breathers Provide shade to reduce sun-scorch.
- It removes harmful metabolites from roots.
- It mobilizes growth substances towards the production of new shoots.
- It contributes photosynthates to the roots.
- It influences the growth of feeder roots.

Few Important Points

- Cut marks should be as small as possible and round in shape.
- Cut marks should always face the center of a bush.
- Height of all the cut marks should be equal from the ground level.
- Cross branches, hanging branches, very thin, whippy branches, branches situated at a very close proximity must be removed.
- Always use measuring sticks at the time of pruning.
- Pruning litters should be preserved *in-situ*.
- Apply Bitumen paint on the bigger cut marks to prevent the infection of *Poria hypobrunnea* .

Light Pruning (LP)

Post Pruning Treatments

(1) Immediately after pruning, spray any systemic fungicide @ 1 gm per litre of water on the cut bushes.

(2) Knife Cleaning out to be followed within a week or so.

(3) The bushes should be completely defoliated.

(4) The big cut marks should be painted with bitumen paint with the help of a brush. Bitumen paint should be heated a bit before application. It should be applied during bright sunshine.

(5) Lime wash of the pruned bush frame helps to clean mosses and lichens. It also helps to kill the eggs of the thrips in the crevices of the tea bushes. The proportion should be as follows :

Quick lime – 1 kg + Caustic soda or washing soda – 3 kg to be dissolved in 50 litres of water, strained and sprayed on the pruned bushes with a hand operated sprayer.

(6) The interspaces should be forked using fork hoe to loosen the soil so that Oxygen enters the soil pore spaces which in turn will promote nitrification process.

(7) Soil correction is done during Light, Medium or lower pruning. If the pH is high, Aluminium Sulphate or Sulphur pyrites should be applied and if the pH is low, Dolomite should be applied on the interspaces after forking. A light irrigation should be given after application.

(8) Organic manuring: In between two bushes, a pit should be dug and following ingredients should be inserted and covered with soil (placement method):

Well dried cowdung / Farm Yard Manure- 1 kg + Rock Phosphate – 30g per pit. FYM / Cowdung should be fortified / enriched using beneficial bacteria *viz Azotobacter, Azospirillum, Bacillus megatherium*, VAM etc.

(9) The drains should be deepened or re graded as per requirement.

(10) Thinning out of the shade trees, if required should be done after Light pruning or pruning lower than that.

(11) The shade tree trunks should be painted with lime upto a height of 2 metres. 10 kg of quick lime + 500 g of Copper oxychloride (or any other fungicide) + 60 ml of Imidacloprid (or any other insecticide) should be dissolved in 200 litres of water and the shade tree trunks should be painted with a brush.

A pruned tea plantation

Medium Pruning (MP) – Big cuts marks pasted with bitumen paint

Deep skiff (DS)

Medium skiff

Deep skiff (DS)

Medium skiff

Light Skiff

Levelling off skiff (LOS)

Breathers

First FFP

Final FFP

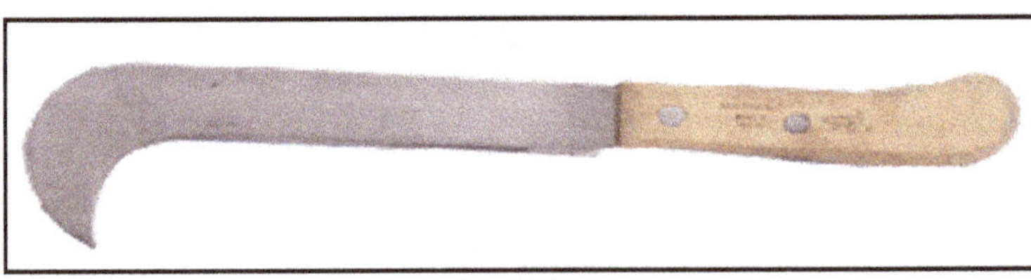

Pruning knife

Secateur

Lime wash of tea bushes

Mechanical Pruning

Looking at the scarcity of workers and increasing cost of manual pruning many garden has resorted to mechanical pruning with the help pruning machines with circular discs.

Pruning Machine

13

PERIODICITY OF GROWTH AND PLUCKING

Periodicity of Growath

Tea plants have a rhythmic growth pattern. Freely growing tea shoots pass through alternating states of growth and dormancy. The amount of growth made between two states of dormancy is known as a 'flush'. In N.E. India, upto four flushes were recorded on unpruned seed trees in a growing season and upto five flushes on pruned bushes, although the fifth flush was rare and found only in a few genotypes.

The apex bud is tightly wrapped in a foliage leaf. Unfolding of the ultimate or 'banjhi' (dormant) buds indicate the beginning of a flush. These 'banjhi' buds are much smaller than growing buds. A 'banjhi' bud is enclosed by two or occasionally three cataphylls or bud scales known as 'janam' (birth scale) in N.E. India. A 'banjhi' bud does not show any visible sign of growth and may remain in that condition for the rest of the season.

Interflush Dormancy

Interflush dormancy occurs in between two flushes i.e. in between 1st flush and 2nd flush, 2nd flush and rain flush.

Causes of Interflush dormancy

(a) It is caused by slow rate of maturation and development of tissues in the growing shoot. That is not able to keep pace with the fast growth and therefore the shoots go dormant,

(b) The shoot also go dormant because of low availability of moisture from the soil and various other factors,

(c) When number of shoots are more in the bush but the production of carbohydrate is less, the shoots go dormant.

Winter Dormancy: Winter dormancy occurs in between autumn flush and first flush.

Causes of winter dormancy

(a) The winter dormancy is caused by the interaction of some environmental factors like short day length and low temperature, which appear to be main factors influencing leaf period.

(b) Tea is cultivated between latitudes 45° north and 35° south, and at or near the equator. It produces almost the same yield every month. But farther from the equator, the winter harvest gradually declines and at latitudes beyond about 16° there is almost complete winter dormancy which lasts longer with increasing latitudes.

(c) Suitable temperature for growing tea is ranges from 12-13^{0}C to 28-32^{0}C. In winter lower temperature comes below 12^{0}C, which affect the hormonal regulation & ultimately causes winter dormancy in tea.

(d) For proper growth and development of tea, day length should not be less than 11 hours 15 minutes. During winter, the day length is short or it is less than 11 hours 15 minutes, hence growth and development of young tea shoots are hampered and shoots go dormant.

(e) Both in winter and inter flush dormancy it has been seen that with the change in the growing pattern of the shoot there is also a change in the endogenous growth promoters/inhibitors ratio. A high ratio makes the shoots grow and a low ratio make the shoots go dormant. According the direction of movement of photosynthates from the maintenance leaf also decided. The growth promoters are auxin, gibberelin, cytokinin and growth inhibitors are abscisic acid and ethylene. The balance is caused by environmental and genetic factors.

(f) Both water deficit and excess water cause moisture stress to the plant. This moisture stress causes morphological, physiological and anatomical changes to the plant which sets dormancy in tea plants.

Spray of Growth Promoters

Spraying growth promoter has not been found effective in preventing the onset of winter dormancy. However, spraying with GA_3 reduces the period of winter dormancy and increases the first flush crop in some clones provided there is no other limiting factor. Thus the degree of response varied depending upon the environmental factors – mainly soil and atmospheric drought,

Application of Antitranspirants

Morphological, physiological and anatomical changes occur in tea plants due to water stress. The ill effect of water stress can be reduced to some extent by foliar application of some anti-transpirants. These chemicals when spread on the canopy conserves moisture and retain turbidity of the shoots. Besides conserving moisture it promotes bud break and flushing after drought. One spray during mid February again in mid March may help the plants to influence the early crop. This also enhance the drought tolerance capacity for about 3 months without affecting the health and productivity of the bushes. 1% MOP(K_20) could also increase water potential, relative turgidity, specific leaf weight and final crop yield in tea.

Dormancy of tea shoots decide the pruning time in tea. It is a good physiological characteristics of tea.

Plucking

Plucking in tea refers to harvesting of tender apical shoots growing above the pre-determined plucking table. Besides giving crop, it encourages regeneration of new shoots, checks vertical growth of bushes and keeps them in vegetative phase. Plucking accounts for 30-33% of total cost of green leaf production.

Objectives of plucking

Young tea – To encourage lateral growth and get some quantity of leaves.

Mature tea –i) To get economic crop for manufacturing. ii) To keep the plants in proper plucking height iii) To reduce dormancy and encourage new shoots iii) To reduce the infestation marks of pests and get more & proper leaf lamina which in turn enhances rate of photosynthesis.

Plucking rounds

Plucking round is the time interval in days between two consecutive rounds. Plucking rounds are ascertained in two ways:

Phyllocron Method: (Leaf period x 2) – 1

Leaf period: Time taken by a 1+ apical bud to become 2 + apical bud, 2+ apical buds to become 3 + apical bud.

Thermal Time Regulated Plucking:

Under N.E. agrological condition, thermal time requirement for buds to become pluckable shoots is 389 day degrees.

$t = A / (T - Tb)$

Where, a) A = Thermal time requirement for buds become pluckable shoots = 389 day degrees.

b) T = Prevailing temperature = (maximum temp. + minimum temp.)/2

c) Tb = Base temperature = 12.50°C

d) t = Time requirement by buds to become pluckable shoots (days)

Standard of Plucking

Black plucking – Removes all the leaves and bud above tipping level except unopened buds.

Standard plucking- In this method all the leaves except apical bud and small one and bud are plucked.

Coarse plucking – One plus apical bud, small two plus apical buds are left and remainder is plucked off.

System of Plucking

Janam plucking – Plucking is done leaving janam on the flush. This is the most common form of plucking in the tea industry.

Fish leaf plucking – Plucking is done leaving janam and fish leaf on the surface. Generally fish leaf plucking is not done in this part. Only during the starting of the season, somewhere it is practiced.

Single leaf or step up plucking – Plucked above a normal serrated leaf. In this system janam, fish leaf and the mother leaf is left on the plucking table. This system is also called mother leaf plucking. After a prolonged drought or after a considerable period of water logging or a huge attack of looper caterpillars and red spider mites when top surface is defoliated due to huge senescence of leaves, this plucking is done.

This plucking is also known as leaf addition. The photosynthetic ability of a tea leaf remains up to 18 months but after 6 months a sharp decline in photosynthesis is noticed. Hence, there is a need to have a new set of leaves with full photosynthetic potential.

Tipping

In a pruned bush, primary shoots arise from the dormant Buds on the pruned sticks. These are decapitated or tipped at pre-determined height to maintain the depth of Maintenance foliage and make a plucking table.

Tipping Heights

LP = 20-26 cm (average height of 5 leaves)

DS = 7-10 cm (average height of 2 leaves)

MS = 4-5 cm (average height of 1 leaf)

LS = at skiffed level (janam plucking)

LOS= at skiffed level (janam plucking)

Collar pruning = 70-75 cm from the ground level

Rejuvenation pruning = 25-35 cm from the ground level

Medium pruning = 25-30 cm from the ground level

Fine Count by Ballometric Method

Fine Count or Ballometric Count is a method which helps us to find the percentage of fine shoot in a flush. Fine leaves are responsible for getting good quality made tea. The fine leaf comprises of two leaves and a bud, small one leaf and a bud and soft single banjhi. It is done by breaking 6mm stem from the bottom of the second leaf.

Flush Component Analysis

Flush component analysis gives information about the status of shoots in a flush. From this it can be ascertained whether plucking round needs to be extended or reduced, whether spray of hormones and growth promoting nutrients to be sprayed or not, whether there is problem in pruning cycle or plucking habit of the pluckers or not etc.

Process

Tender tea shoots are plucked and collected.

Shoots with active buds and shoots with dormant buds are selected.

From there 1+b, 2+b, 3+b, 4+b, 1+d, 2+d, 3+d, 4+ d single leaves are separated.

Weight of 1+b is measured. By applying same method number and weight of other shoots are determined.

	Active shoots					Dormant (Banjhi) shoots			
	No.	%	Weight	%		No.	%	Weight	%
1+ bud					1+ banjhi				
2 + bud					1+ banjhi				
3 + bud					1+ banjhi				
4 + bud					1+ banjhi				
Soft single banjhi					Hard single leaf				

When a leaf is folded, if the bud is longer than half the leaf, it is an active shoot. If the bud is shorter than half of the leaf it is a dormant (banjhi) shoot.

Looking at the flush component analysis, the status of the plucked leaves is determined. On the basis of the status, root cause analysis is done and corrective action is initiated.

a) If the percentage of dormant shoot is more then it will indicate that the plucking round is extended or there may be hormonal problem or it may be the effect of bad plucking or adverse weather condition.

b) If the percentage of 1 + bud is more then it will indicate that the plucking round is too short.

c) If the percentage of hard single leaf is more, it will indicate that there is problem in plucking method.

d) If the percentage of 3 + bud is more, it will be considered as a perfect plucking under Dooars and Terai conditions.

How to Improve Pluckers' Productivity ?

High Yielding Sections: It has been observed that in high yielding sections productivity of the pluckers is more Hence, proper seed and clone stock should be selected for plantation in accordance with agro-ecological conditions.

LEAF SIZE : Pluckers productivity would be more in the sections having tea plants with more leaf area. For example TV(Tocklai Vegetative) 30 has bigger leaf area and it is a high yielding clone.

Height of the Plants : Height of the plants should match the height of the pluckers. Hence, Anthropology has got something to do with the height of the tea plants. Proper pruning programme and control on creep can help. Longer plucking rounds also induces unnecessary creep. Hence, fix plucking rounds as per phylocron system.

Free Passage for the Pluckers : There should be no obstruction on the way of the pluckers. Weeds, hanging branches and congested bush frames reduce pluckers productivity. Hence, it should be endeavoured to keep the section weed free. Hanging branches, if any should be trimmed as and when necessary. Lane cutting may also be practiced. Single hedge plantation also helps.

Categorisation of Pluckers on the Basis of Plucking : There are three types of pluckers – fast, medium and slow. Fast pluckers pluck 1/3 rd more than medium and slow pluckers pluck 1/3 rd less per unit time. If fast and slow pluckers pluck together,

the slow pluckers ten to leave shoots on the bush surface to catch up with the fast pluckers. Hence deployment category should be taken care of.

Distance Between Plucking area to Weighment Spot : Arrangement should be made to take weighment in the field itself very near to the plucking area, so that walking time is reduced.

Less Waiting Time : More spring balances should be employed so that waiting time for the shake of weighment can be minimized.

Less Leaf Carriage : By planning frequent weighment (say thrice a day), the burden of the pluckers may be lessened.Where three times weighment is not possible till the factory is installed, each plucker should be provided with at least two carrying bags (net / hessian)

Role of Shade Trees : During droughty summer days, the role of shade trees in improving pluckers' productivity is immense. Right kind of shade trees should be planted. It is imperative to plant shade trees which posses leaves during March – April. *Albizzia odoratisima* and *Derris robusta* would be the best choice. Shades like *Albizzia chinensis, Acacia lenticularis, or Albizzia lebbek* whose leafless period fall during March –April should be avoided.

Standard Of Plucking : The pluckers should be taught about the quality and standard of plucking. The plucking round should be so fixed that the leaves received at the factory should bear following proportions:

3 + bud / banjhi (dormant) shoots	:	90 %
Course leaf	:	maximum 5%
Immature buds	:	maximum 5%

Creche , proper arrangement for drinking water during plucking time and light music have been found effective in improving pluckers' productivity

Sticks: Both in pruned and unpruned sections, pluckers should use measuring " T " sticks. For prune / D.S. / M.S. sections a Tee may be prepared with extended arms to cover the entire plucking surface.

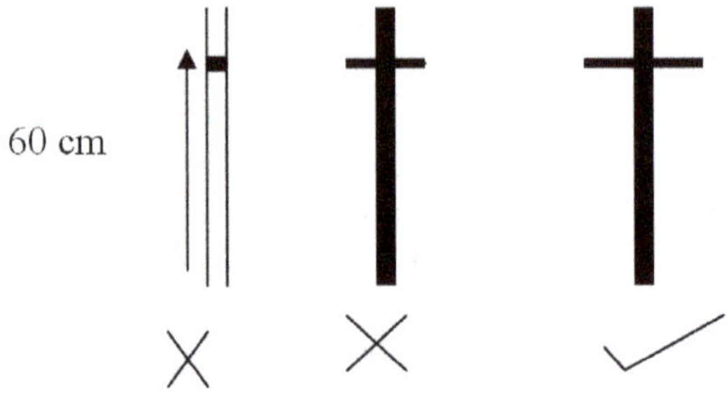

In skiffed sections a stick may be placed horizontally on the top surface so that the plucker can check their habit of dipping hands or pluck lowly placed side branches

Mother Leaf Plucking and Level Plucking: The plucker should be taught that the maintenance foliage which is the source of photosynthesis can retain peak photosynthetic ability maximum upto 6 months. Hence during the initial parts of the first flush and end of the season, mother leaf plucking should be practiced, while during peak growing season level plucking will help.

Training of the Supervisors: Training of the supervisors involved in plucking operation should be trained properly by AUDIO – VISUAL devices .They may also be sent to gardens with high profile of plucking .

Plucking Shears: To control plucking rounds and evenness of the surface during peak season (heavy flush), plucking shears may be introduced. Flat shears (30 cm length) should be used – not the two step shears.

Plucking Task : The plucking task must be fixed very rationally, so that to fulfill the task the pluckers are not compelled to destroy the physiology of the tea plants

Incentive: It has been observed that rational increase in extra – leaf price improves pluckers productivity.

In future the concept of THERMAL TIME REGULATED PLUCKING may be introduced.

Mechanical Harvesting

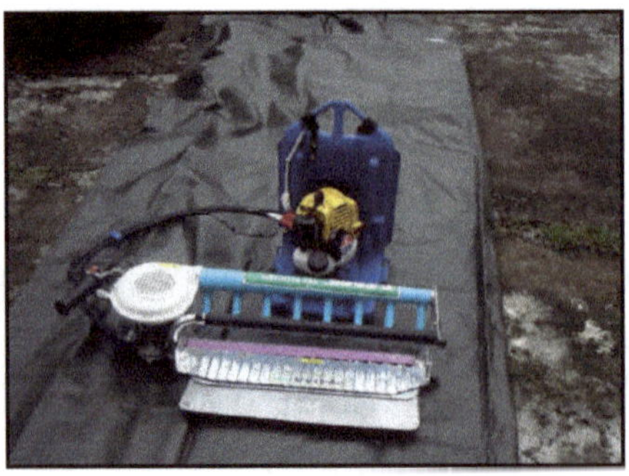

A Plucking Machine

In India about 67 % (about two-third) of the plantation workers are involved in plucking and about 33 % (about one-third) of the total expenditure in a tea plantation goes to plucking operation alone. In recent years there has been a sharp decline in the percentage of young workers in the tea plantation because the new generation off springs of the plantation workers are opting for career other than tea. In South India mechanical plucking is quite significant but in North East India, mechanical harvesting is yet to make in roads. To cope up with the crisis, it has become essential to introduce mechanical plucking in the tea industry. Mechanical plucking is also considered to be cost effective. Statistics on manual plucking and mechanical plucking is furnished below :

	1-man operated machine	**2-men operated machine**	**Hand shears**	**Manual**
Plucking Avg (kg) /worker	180	150	60	40
Total Plucking per day (kg)	450	750	60	40
Worker /hectare	6	8.33	15	20
Area Covered per day(hectare)	0.3-0.5	0.5-0.75	0.07	0.05
No. of machine /hectare	2-3	1.3-2	-	-
Plucking interval(days)	18-22	22-25	12-14	8-10

Cost Economics/ Hectare

	1-man operated machine	**2-men operated machine**	**Hand shears**	**Manual**
Labour cost(Rs)	512.5	1025	205	195
Fuel cost (Rs)	180	252	0	0
Maintenance cost (Rs)	350	600	2	0
Cost/kg	2.36	2.53	3.47	4.88

Note : Above statistics are generated by Tea Research Association, Tocklai Experimental Station.

14

PEST CONTROL

Definition of Pests

1. Pests are any form of plant or animal life that is injurious or potentially injurious to human being and its economy.
2. A plant or animal detrimental to humans or human concerns (as agriculture or livestock production).
3. Example: insects, vertebrates, weeds, pathogens such as bacteria, fungi, viruses.

Insects

1. Usually small arthropod animals of the class Insecta.
2. Having an adult stage characterized by three pairs of legs and a body segented into head, thorax, and abdomen and usually having two pairs of wings.
3. Insects include the flies, crickets, mosquitoes, beetles, butterflies and bees.

Mites

1. A tiny four or eight-legged creature belonging to the Order Acarina related to spiders.
2. Some mites live freely and others as parasites. Mites can attack plants and animals, carry disease, and cause allergies.

Classification of Pests

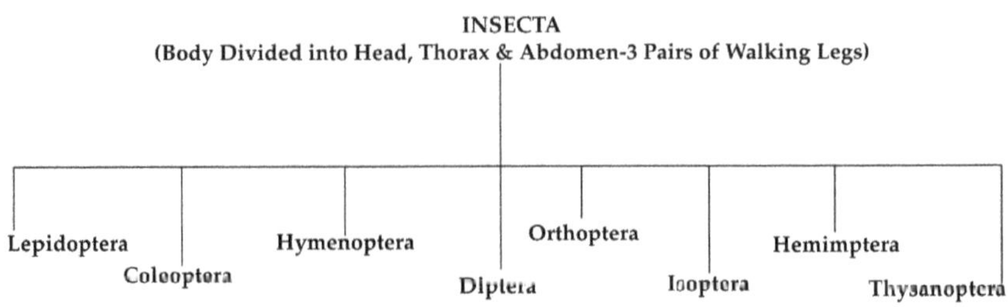

Lepidoptera – Lepidos = Scale ; Pteron = Wing

Examples : Leaf eating caterpillars eg Looper caterpillars, Red Slug caterpillars, Bunch caterpillars, Flush worm, Leaf roller, Lymantrid caterpillars

Coleoptera - Koleos = Sheath; Pteron = Wing

Examples : Cock chafer grub, Leaf eating beetle

Hymenoptera – Hymen = membrane; Pteron = Wing

Examples : Sewing blight wasp, ants

Diptera – Dis= two; Pteron = Wing

Example : Tea leaf miner

Orthoptera – Orthos = Straight; Pteron = Wing

Examples : Grass hopper, Locust, Cricket

Isoptera – Isos = equal; Pteron = Wing

Example : Termites

Hemiptera – Hemi = Half; Pteron = Wing

Examples : Tea mosquito bug, Mealy bug

Thysanoptera – Thysanos = Fringed; Pteron = Wing

Example : Thrips

Classification of Mites

Mites (Acarina)
(Body divided into 2 parts-Gnathosoma & Podosoma; 2 or 4 pairs of Walking Legs)

- **Tetranychidae**
- **Tenuipalpidae**
- **Eriophydae**

Tetranychidae – 4 pairs of legs, long recurved stylet, generally found on the dorsalsuface of the leaves

Example: Red spider mite

Tenuipalpidae- Scarlet black marking on the dorsal side of the body, generally found on the ventral surface of the leaves.

Example : Scarlet mite

Eriophydae – 2 pairs of legs, body elongated, worm like

Examples: Pink mite and purple mite

Principles of Pest Control

- Identify the pest.
- Know what control measures are available.
- Evaluate the benefits and risks of each method or combination of methods.
- Recognize when direct action such as pesticide application is necessary.

- Choose the methods that are most effective and eco-friendly.
- Know the correct use of control.
- Obtain knowledge about Economic Threshold Level (ETL) and Economic Injury Level (EIL)

Root Damaging Pests

Root Knot Nematodes (EEL Worms)

Root Knot Nematodes are commonly known as eelworm (Order: Nematode).

The pathogenic nematodes are of two types:

1. Root-Knot Nematodes,
2. Root-lesion nematodes

Harmful species. of Root-Knot Nematodes are:

(a) Meloidogyne incognita

(b) Meloidogyne javanica

(c) Meloidogyne hapla

(d) Meloidogyne brevicaudata

Harmful species of Root-lesion Nematodes are:

(i) Pratylenchus brachyurus

(ii) Pratylenchus loosi

Damage Symptoms of Nematodes

Root-Knot Nematodes

1. Usually attack young seedlings and clonal plants in nursery and sometimes in the field during the first year of planting.
2. The infected roots exhibit numerous galls or knots, swelling, curved and twisted formations .
3. These changes interfere with the uptake of water and nutrients as a result plants suffer from drought.

Root Lesion Nematodes

1. They feed and lay eggs within the cortex of young roots and move from root to root in search of food.
2. These movements and feeding activity disintegrate cortical tissues and retard growth of the plant.
3. Such debilitated plants may not recover after pruning.

Few Facts and Control Measures

1. Care should be taken in the selection of nursery sites since nurseries play an important role in the spread of nematodes.
2. Virgin land or land under wild vegetation for a number of years should be selected for nursery.
3. Soil considered unsuitable for nursery on account of high eelworm count can be rehabilitated by putting under green crop *Crotolaria anagyroides* at least for two years. *Crotolaria* is not infested by eelworm.

4. It should not be above 4 nos./10 g. of soil. If it is more it can be controlled by the following methods :
   (a) Calculate the quantity of soil required to fill 1000 polythene sleeves. In that quantity of soil, add 1000 g of carbofuran thoroughly or 2 g. carbofuran / sleeve,
   (b) Uniform heating (after sieving twice through GI wire mesh No. 4 for eliminating any vegetative matter) of the soil up to 60-70^{0}c for 4-5 minutes on plain tin sheets.

Cockchafer

Cockchafer grub

Cockchafer beetle- adult

Out of several species of cockchafer grub particularly *Holotrichia impressa* (Order: Coleoptera) is the most destructive in Dooars and Terai.

Identification

Eggs are oblong and glistering white in colour,

Grub whitish in colour,

Pupae yellowish in colour (20-22 mm in length),

Adult beetles are deep brown in colour.

Site of attack

(a) It generally found in nursery as well as in field up to 3 years after transplanting

(b) The grub attack collar region of young plant (below 3 years) and seedling & roots of young tea,

(c) The adult cause damage to foliage of shade trees.

Damage symptoms

1. Larvae eat away the bark in the collar region just below the soil surface either in a ring or in patches.
2. The damage done by cockchafer is very similar to that of manure damage.

Life-cycle

Site of oviposition:

Eggs are laid in the soil at a depth of 4-5 cm.

Fecundity: 40 eggs

Incubation Period: 10-11 days

Peak period of infestation: June – August (grub) During April beetles settle down on the trunk of shade trees, fencing etc. where from these can be hand collected and destroyed.

Control Measures

Bio-pesticide: *Metarhizium anisoplae*: 800 g/ drum

Synthetic chemicals:

1. Imidacloprid: 80-100 ml/drum
2. Chlorpyriphos :300ml/drum

Termites

Species of Termite

1. Live wood eating termites: *Microcerotermes* sp.
2. Scavenging termites: *Odontotermes assamensis.*

Seasonal occurrence

Initial infestation starts:– August – October

Peak:- November – February (Winter)

Site of attack: Trunk, branches and stems.

Kinds of Termites

(a) Live wood eating termites form many small nests of 8 – 20 cm below in the soil.

(b) Queens of termites produce eggs every second, hence must be killed for effective control.

Damage

1. The scavenger first attack the dead wood resulting from fungal or borer attack, also extended the damage to the adjacent healthy tissues by removing the dead woods.
2. The live wood-eating termites attack the tissue of the bark of healthy young bushes. They excavate fine tunnels into the healthy wood and hollow out the stem leaving only a thin layer of wood and bark or whole woody part with earth materials leaving the bark out side.

Control Measures

Cultural:

(i) Clean the infected bush thoroughly at pruning and put indopaste at the cut ends.

(ii) Improve drainage and shade status.

(iii) Destroy termite mound and queens.

Bio-Pesticide:

Metarhizium anisoplae : Dose: 800 g/ drum.

Synthetic Chemicals:

Imidacloprid: 80 – 100 ml/ drum

Chlorpyriphos (Dursban): 300 ml/ drum

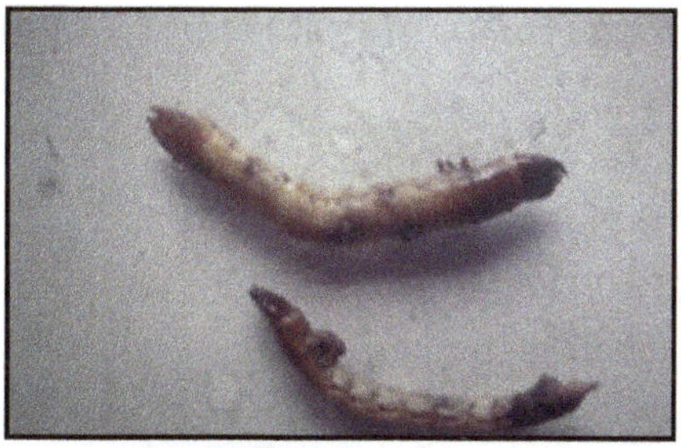

Red Borer

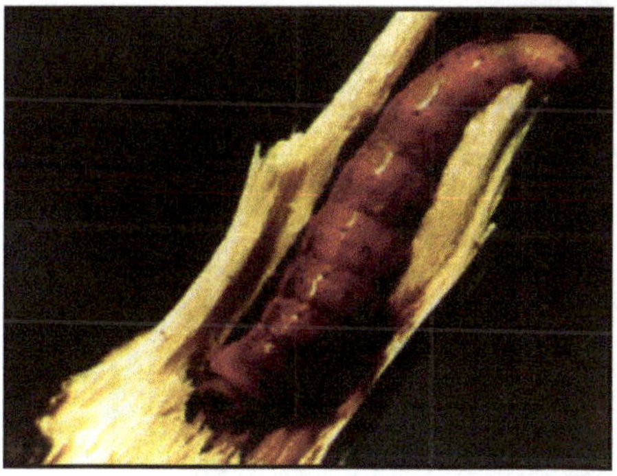

Moth of red borer (moth)

Red Borer (Zeuzera coffeae)

1. Damaging Stage:- Larva

2. Adult

Orange coloured, medium sized moth.

The adult has white and black spotted wings.

The larva is red coloured.

3. Life-cycle

Eggs are usually laid singly on the bark.

Eggs hatch in 8 – 12 days and larval development takes 3 – 4 months.

Pupation takes place inside the stem and the adult emerges in about 2 months.

Adult is a moth.

Damage Symptoms

Caterpillar bores into the stem or branches to feeds on the wood.

In early stages of attack, young plants or braches show wilting.

Pellet-like excrement of the larva hangs out and accumulates at the base of the plant.

In advanced cases, the branch dries up.

Control Measures

Affected plants or twigs should be cut and burnt.

Providing good shade to plantation.

Field release of white muscardine fungus *Beauveria bassiana* and parasite *Amyosoma zeuzerae.*

Stem Borer *Xyleborus fornicates (Order: Coleoptera)*

Damaging stage : Adult

Life Cycle

Incubation- 10-12 days

Larval period- 45-60days

Pupal period- 20-30 days (tunnel)

Activity- Winter months

Major Sucking Pests

1. Tea Mosquito bug : *Helopeltis theivora*
2. Tea Thrips : *Scirtothrips dorsalis*
3. Tea Jassids : *Empoasca flavescens*
4. Aphis : *Toxoptera aurantii*

Tea mosquito bugs on the leaf surface

Tea mosquito bug punches

Leaf crippled after severe attack of Tea mosquito bug

Tea Mosquito Bug

Damaging stage – Nymphs and Adults

Seasonal incidence

I. Feb – Mar:- Building up of the population
II. Apr – May:- Infesting tea leaves
III. June – Sept:- attaining peak
IV. Oct – Mid Dec:- Retaining of population
V. January:- dwindles down gradually

Site of attack

The nymphs and adults suck the sap of young leaves, buds and tender stems, Growing shoots and bud at the leaf axils are the oviposition sites.

Life Cycle Of TMB

Adult lays eggs on midrib of leaves

Fecundity - A bug may lay 4-5 eggs/day. A single female may lay more than 200 eggs.

Incubation period- 10-13 days.

Life cycle of *Helopeltis* completed in 10-35 days.

Nymphal period- 22-23 days.

Adult - Black head, pale yellow and black thorax, yellow and greenish black abdomen.

Nymph is dirty yellow.

Damage Symptoms

1. Tea leaves are damaged by the nymphs and adults.
2. Initially, a ring like spot is formed at the sap-sucking point turning translucent light brown within 24 hours.
3. These spots later on turn dark brown and appear as sunken spots and dry up.
4. An adult can produce about 100 such spots within 24 hours.

Tea Thrips (*Scirtothrips dorsalis*) (Order: Thysanoptera)

Tea thrips

Sand papery lines on the leaf surface

Damaging stage – Nymphs and Adults

Seasonal incidence: March – April: Rapid Build up

May: Attaining peak

Life-cycle

Fecundity: Eggs are laid singly, completely embedded in the tissue of leaf buds and young leaves.

Incubation period: 3 – 4 days.

Larva: 1 – 3 days.

Pupa: 3 days. Pupation takes place in lichens and mosses growing on tea bushes.

Total duration of life cycle: 14 – 20 days.

Damage Symptoms

1. Leaf surface becomes uneven, curly and matty, exhibiting parallel lines of feeding marks on either side of the midrib. Puncture marks appear as minute brown spots, later on 'Sand papery' lines (2 - 4) appear on either side of the midrib.
2. Heavily infested fields sometimes acquire a bronze colour.
3. Badly damaged leaves appear rough, deformed and curled.

Tea Jassids : *Empoasca flavescens* (Order : Hemiptera)

Tea Jassid (Grren Fly)

Damaging stage- Nymphs and Adults

Seasonal incidence

(a) May – July: The pest is more active

(b) August onward: Population suddenly declines to a negligible number.

Life-cycle

1. Fecundity: 100 eggs during her lifetime.
2. Depending on the temperature, the eggs will hatch after 6-13 days.
3. Eggs are laid by inserting in the tissue of new tea buds and shoot, particularly in the internode. In one bud up to 7 eggs are laid.
4. Nymphal period: Total nymphal development period is 7 - 16 days being faster in warm weather condition. There are five nymph stages.
5. Total duration : The life cycle completed in 14 – 28 during March-July.

Damage Symptoms

1. Both adult and nymphs, suck the sap of young leaves, cause damage, nymphs are responsible for greater damage than the adults.
2. The affected leaves become uneven ad the leaves curl downward, the margin become recurved and subsequently turns brown and dry up. This characteristic symptom is known as the rim blight.

The mid rib and veins of the affected leaves may also show somewhat brownish discolouration.

Jassid infested tea plant

Aphis – *Toxoptera aurantii* (Order : Hemiptera)

Damaging stage: Nymphs and adults.

Site of attack: Buds, Tender stems, Young leaves.

Seasonal incidence: January – April.

Damage symptoms: Crinkled and curled leaves below tipping level.

Control Measures

Cultural Control

1. Thinning out of shade trees in heavy shaded areas.
2. Check the egg population either by hard plucking or level of skiff in heavily infested section.
3. Prune/ deep skiff heavily infested sections by *Helopeltis* in the previous season.
4. Remove weed or other alternate hosts. Alternate weed hosts for *Helopeltis* are *Melastoma malabathricum, Mikania micrantha, Oxalis acetocella*.
5. Alkaline/ lime/ caustic washing of the trunk of the bushes after cleaning the mosses and lichens and stirring of soil around the collar region will kill the pupae of thrips during Dec. - March. Lime 1 kg + Caustic soda or washing soda @ 3 kg in 50 litres of water.
6. Ladybeetles, praying mantids feed on jassids.
7. Tea growing under shade trees tends to have less jassid problem. Mulching also helps to increase humidity and therefore reduce jassid populations.
8. Use of light taps.
9. Sex-Pheromone traps for example: Sticky Delta Trap, Wota Trap, Fero Trap.

Mechanical and Physical Control:

1. Hand collection of the adults & nymphs of Tea Mosquito bug – the adults and nymphs are active during early morning, late afternoon and night on the top hamper of bushes.
2. Use of yellow colour sticky traps for monitoring of thrips.

Bio-Pesticide:

1. *Beauveria bassiana*:- Dose: (i) Liquid 200 ml/ drum
   (ii) Powder: 600 g/ drum
2. *Verticillium lecanii*:- Dose: (i) Liquid 200 ml/ drum
   (ii) Powder: 800 g/ drum

Leaf Eating Pests

Characteristics of Lepidopteran Insects

Lepidopteran species are characterized by more than three derived features, some of the most apparent being the scales covering their bodies and wings, and a proboscis. The scales are modified, flattened "hairs", and give butterflies and moths their extraordinary variety of colours and patterns. Almost all species have some form of membranous wings, except for a few that have reduced wings or are wingless. Like most other insects, butterflies and moths are holometabolous, which means they undergo complete metamorphosis. Mating and the laying of eggs are carried out by adults, normally near or on host plants for the larvae. The larvae are commonly called caterpillars, and are completely different from their adult moth or butterfly form, having a cylindrical body with a well-developed head, mandible mouth parts, and from 0 to 11 (usually 8) pairs of prolegs. As they grow, these larvae will change in appearance, going through a series of stages called instars. Once fully matured, the larva develops into a pupa, referred to as a chrysalis in the case of butterflies and a cocoon in the case of moths.

External Morphology

- Lepidoptera are morphologically distinguished from other orders principally by the presence of scales on the external parts of the body and appendages, especially the wings.
- Lepidopterans undergo a four-stage life cycle: egg; larva or caterpillar; pupa or chrysalis; and imago/ adult and show many variations of the basic body structure, which have evolved to gain advantages in lifestyle and distribution.
- Body is divided mainly into three parts - head, thorax and abdomen.

Head

The head is where many sensing organs and the mouth parts are found. Like the adult, the larva also have a toughened, or sclerotized head capsule. There are two compound eyes, and chaetosema, raised spots or clusters of sensory bristles unique to Lepidoptera, even though many taxa have lost one or both of these spots. The larvae, called caterpillars, have a toughened head capsule. Caterpillars lack the proboscis and have separate chewing mouthparts. These mouthparts, called mandibles, are used to chew up the plant matter that the larvae eat. The lower jaw, or labium. The head is made of large lateral lobes, each having an ellipse of up to six simple eyes.

Thorax

The thorax is made of three fused segments - the prothorax, mesothorax, and metathorax,each with a pair of legs. The first segment contains the first pair of legs. The three pairs of legs are covered with scales. Lepidoptera also have olfactory organs on their feet, which aid the butterfly in "tasting" or "smelling" out its food. The two pairs of wings are found on the middle and third segment, or mesothorax and metathorax respectively. The caterpillar has an elongated soft body that may have hair-like or other projections, 3 pairs of true legs, with 0–11 pairs of abdominal legs (usually 8) and hooklets, called apical crochets. The thorax will usually have a pair of legs on each segment. The thorax is also lined with many spiracles on both the mesothorax and metathorax, except for a few aquatic species, which instead have a form of gill.

Abdomen

The abdomen, which is less sclerotized than the thorax, consists of 10 segments with membranes in between allowing for articulated movement. The sternum, on the first segment, is small in some families and is completely absent in others. The last 2 or 3 segments form the external parts of the species' sex organs. The abdomen of the caterpillar has 4 pairs of prolegs, normally located on the third to sixth segments of the abdomen, and a separate pair of prolegs by the anus, which has a pair of tiny hooks called crotchets. These aid in gripping and walking, especially in species that lack many prolegs.

Scales

The wings, head, parts of the thorax and abdomen of Lepidoptera are covered with minute scales, a feature from which the order 'Lepidoptera' derives its name. Most scales are lamellar, or blade-like and attached with a pedicel, while other forms may be hair-like or specialized as secondary sexual characteristics.

The lumen or surface of the lamella has a complex structure. It gives colour either by coloured pigents that it contains, or through structural coloration with mechanisms that include photonic crystals and diffraction gratings.

Caterpillars

Looper Caterpillar: *Biston (Buzura) suppressaria, Hyposidra talaca, Hyposidra infixaria, Ectropis* sp. , *Ascotis* sp., *Cleora* sp.

Red slug caterpillar: *Eterusia magnifica*

Bunch Caterpillar: *Andraca bipunctata*

Looper Caterpillars

1. Time of occurrence

   a. Caterpillar: March - October

   b. Pupa: April – November

   c. Moth: Feb – Mar/May/Jul/Aug - Oct

2. Eggs and life cycle

   a. Bluish green covered with buff coloured hairs, laid on shade tree trunk.

   b. Hatch in 8 – 9 days into black caterpillars.

   c. Larval stage: 24 – 36 days

   d. Pupation: 20 – 22 days in soil at 5 – 10 cm Deep

Looper caterpillar damage

Different Types of Looper Caterpillars

(1) Biston (=Buzura) suppressaria

Moth of *Biston (=Buzura) suppressaria*

Larva of *Biston (=Buzura) suppressaria*

This is the earliest known species of looper caterpillar infesting tea. The moth is greyish white, finely speckled with black. The males are generally smaller than the females and are easily distinguished by the bipectinate (feathered) antennae. The eggs are laid in heaps, each containing 200-600 eggs covered with a buff coloured hairs. The most common site for the deposition of eggs is the trunk of shade trees or any other tall trees in the vicinity of tea area. The young caterpillar is dark brown with greenish white lines along the back and side. The body colour soon turns to light green and with age, it acquires a brownish grey colour similar to that of a mature twig of tea.

(2) *Hyposidra talaca*

Moth of *Hyposidra talaca*

Larva of *Hyposidra talaca*

Early instars caterpillars are brown to black in colour with transverse six white bands on the dorsal side of the body. In mature larva the bands are disappeared and change its colour to brownish grey with white spots on the body. The full grown caterpillar is measured about 45-50 mm in length. Adult moths are greyish to light brown in colour and the females are much larger than the male with a bulging abdomen. The caterpillar passes through 5-6 larval instars. Pupation generally takes place in soil around the collar region of tea bushes and shade trees. In old tea bushes, it prefers the cracks and crevices of the frame to pupate.

(3) *Hyposidra infixaria*

Early instars caterpillars are brown to black in colour with transverse six white bands on the dorsal side of the body. The colour of mature larva is brownish grey with two transverse lines at the thoracic region of the body. The adults are pale or greyish in colour. The distinguishing character of this species is that the male moths bear a tuft of hairs at the abdominal tip.

Moth of *Hyposidra infixaria*

Larva of *Hyposidra infixaria*

Moth of *Ectropis* sp.

Larva of *Ectropis* sp.

***(4) Ectoropis* sp.**

The early instar caterpillars are brownish in colour with black stripes on the dorsal side of the body.The full grown caterpillar is about 30- 35 mm in length and look like a dried up twig of tea bush.

The adults are pale yellow in colour with black wavy lines. The males are smaller than the female with bipectinate antennae.

Moth of *Ascotis* sp.

Larva of *Ascotis* sp.

(5) *Ascotis* sp.

The early instar caterpillar is light green in colour. A matured caterpillar changes its colour to greenish grey with a pair of tubercle located dorsally on the abdominal region of the body.

The adults are greyish green in colour with brown spots on each wing.

Moth of *Cleora sp.*

Larva of *Cleora* sp.

(6) *Cleoro* sp.

The wing base of moth is white with creamy to black or even greyish black. Unlike *Ascotis* sp, the forewings are somewhat rounded. Fully grown caterpillars are robust with orange head and thoracic legs.

There is a pair of reddish brown tubercles dorsally and a smaller pair dorso-laterally in anterior abdominal segment of full grown caterpillar with a narrow transverse greyish bar anterior to them.

Damage Symptoms

The damage symptoms of different species are more or less similar. The newly hatched caterpillars make small holes along the margins of young leaves and then eat from the margins. Young ones are very often seen on growing shoots and buds scrapping on the surface also. The grown up caterpillars prefer semimature to mature leaves. From late 3rd instars, they become voracious feeder. In severity, bushes are completely stripped of the foliage. Even bark of small branches is also eaten away in absence of foliage.

Red Slug Caterpillars

Red Slug caterpillar

Leaf infested by Red slug caterpillar

1. Time of occurrence:

a. Caterpillar: Mar – Apr/May – Jun/Jul – Aug/Sep - Oct.

b. Moth: Feb – Mar/May/Jul/Aug - Oct.

2. Eggs and life cycle:

a. Oval, pale yellow, under surface of leaves & stems.

b. Hatch in 8 – 12 days in May – June.

c. Larval & Pupal stages: 4 – 5 and 2 -3 weeks respectively.

d. Life cycle: 8 – 10 weeks in summer months.

3. Damage symptoms

Firstly feeding on epidermal tissues and skeletonizing the lower mature leaves of the bush.

Thereafter migrating to the top hamper and feeding on leaf margins.

Leaf defoliation and bark damage in severe infestation.

Bunch Caterpillars

1. Time of occurrence:

a. Caterpillar: March – Apr/May/ - Jun/Jul – Aug/Oct - Nov

b. Pupa: April – November

c. Moth: Mar –Apr/May – Jun/Jul – Aug/Oct – Nov.

2. Eggs and life cycle:

a. Oval, yellow on under surface in rows.
b. Hatch in 10 – 11 days in summer
c. 8 – 9 weeks in march – June.

3. Damage symptoms:

a. Young ones feeds on epidermal tissues and margins.
b. Later Instars caterpillars devour entire leaves, remains in cluster during the day and highly active after sunset.

Control Measures

Cultural Control

1. Marking out of infested sections.
2. Collection and destruction of chrysalides:
   i. Looper: Dec. – Feb. : Soil
   ii. Red Slug: Nov. – Dec.: Dry leaves/ pruning litter etc.
   iii. Bunch: Nov. – Jan.: Dried Leaves
3. Trapping of moths by hand collection or light trapping:
   i. Looper: Feb. – Mar. and Aug. - Oct.
   ii. Red Slug: Nov. – Jan. and Feb. - Apr.
   iii. Bunch: Oct. – Nov. and Mar. – Apr.

Bio Pestcides

Bacillus thuringiensis (Hatt): Dose: 800 g/ drum.
Nuclear Polyhedro Virus (NPV): 200 ml/ drum.

Plant Extract

Karanja extract: Dose: 200 g./ drum

Pesticides

List of pesticides, which are available in the Indian market is furnished below as a ready reckoner for the tea planters. However, it is advised that only Tea Board approved pesticides should be used in the tea plantations.

Tea Board Approved Insecticides (Plant Protection Code)

Bifenthrin 8 % SC
Deltamethrin 2.8% EC
Enamectin Benzoate-5% SG
Fenpropathrin 30% EC
Flubendiamide 39.35% SC
Phosalone 35%EC
Deltamethrin 2.8%EC
Profenophos 50% EC
Quinalphos 25% EC & 20% AF
Thiacloprid 21.7% SC
Thiamethoxam 25% WG
Clothianidin 50% WDG

NAMES OF A FEW INSECTICIDES AVAILABLE IN THE MARKET (2014)

	Trade Name	Company	Dose	Target Pest
Acephate 25% + Fenvelarate 3 % EC	Lucid Power	Cheminova India Ltd.	3 ml/litre	Sucking & chewing pests
Acephate 50% SP + Imidacloprid 1.8% SP	Lancergold	United Phosphorus Ltd.	2.5 g/litre	Sucking pests mainly
Acephate 75% SP	Torpedo	Isagro(Asia) Agrochemicals Pvt. Ltd.	2.5 g/litre	Sucking pests
	Sritaf	Crystal Phosphates Ltd.	2.5 g/litre	Sucking pests
	Acevip	Godrej Agrovet Ltd.	2.5 g/litre	Sucking pests
	Lucid	Cheminova India Ltd.	2.5 g/litre	Sucking pests
	Hilphate	Hindustan InsecticidesLtd.	2.5 g/litre	Sucking pests
	Tremor	Biostadt India Ltd.	2.5 g/litre	Sucking pests
	Acefex	Excel Crop Care Ltd.	2.5 g/litre	Sucking pests
	Pace	Nagarjuna Agrichem Ltd.	2.5 g/litre	Sucking pests
	Dhanraj	Dhanuka Agritech Ltd.	2.5 g/litre	Sucking pests
	Oval	PI Industries Ltd.	2.5 g/litre	Sucking pests
	Asataf	Rallis India Ltd.	2.5 g/litre	Sucking pests
	Chetak	GSP Crop Science Pvt. Ltd.	2.5 g/litre	Sucking pests
	Rythane	SDS Ramcides Crop Science Pvt. Ltd.	2.5 g/litre	Sucking pests
	Starthene	Swal Corporation.	2.5 g/litre	Sucking pests
Acetamiprid 20% SP	Ekka	Krishi rasayan	0.3 g/litre	Sucking pests
	Dhanpreet	Dhanuka Agritech Ltd.	0.3 g/litre	Sucking pests
	Opal	Chemtura Chemicals India Pvt. Ltd.	0.3 g/litre	Sucking pests
	Acetamiprid 20% SP	Makhteshim – Agan India Pvt. Ltd.	0.3 g/litre	Sucking pests
	Cilking	Cheminova India Ltd.	0.3 g/litre	Sucking pests

Contd....

	Trade Name	Company	Dose	Target Pest
	Pride	Dow Agrosciences India Pvt. Ltd.	0.3 g/litre	Sucking pests
	Inspector	Godrej Agrovet Ltd.	0.3 g/litre	Sucking pests
	Ennova	Nagarjuna Agrichem Ltd.	0.3 g/litre	Sucking pests
	Theme	SDS Ramcides Crop Science Pvt. Ltd.	0.3 g/litre	Sucking pests
	Nize	GSP Crop Science Pvt. Ltd.	0.3 g/litre	Sucking pests
	Rapid	Crystal Phosphates Ltd.	0.3 g/litre	Sucking pests
	Tata Manik	Rallis India Ltd.	0.3 g/litre	Sucking pests
	Excel Acetacel	Excel Crop Care Ltd.	0.3 g/litre	Sucking pests
	Permit	Isagro(Asia) Agrochemicals Pvt. Ltd.	0.3 g/litre	Sucking pests
Alphamethrin 10%	Merit Alpha	Chemtura Chemicals India Pvt. Ltd.	0.5 ml/litre	Chewing pests
	Alpha Super	Crystal Phosphates Ltd.	0.5 ml/litre	Chewing pests
	Farmex	Excel Crop Care Ltd.	0.5 ml/litre	Chewing pests
	Forward	Isagro(Asia) Agrochemicals Pvt. Ltd.	0.5 ml/litre	Chewing pests
	Dolfin	GSP Crop Science Pvt. Ltd.	0.5 ml/litre	Chewing pests
	Gem	Indofil Industries Ltd.	0.5 ml/litre	Chewing pests
	Stop	Biostadt India Ltd.	0.5 ml/litre	Chewing pests
	Rodio	PI Industries Ltd.	0.5 ml/litre	Chewing pests
	Numethrin	Cheminova India Ltd.	0.5 ml/litre	Chewing pests
	Guru	United Phosphorus Ltd.	0.5 ml/litre	Chewing pests
	Tata Alpha	Rallis India Ltd.	0.5 ml/litre	Chewing pests
	Alpha	Krishi Rasayan	0.5 ml/litre	Chewing pests
Betacyclofluthrion 9 % + Imidacloprid 21%	Solomon	Bayer Crop Science Ltd.	1 ml/litre	Sucking pests
Bifenthrin 10 % EC	Bisect	United Phosphorus Ltd.	1 ml/litre	Sucking & chewing pests
	Markar	Dhanuka Agritech Ltd.	1 ml/litre	Sucking & chewing pests

Contd...

	Trade Name	Company	Dose	Target Pest
	Imperial	Makhteshim – Agan India Pvt. Ltd.	1 ml/litre	Sucking & chewing pests
Buprofezin 25% SC	PI Bupro	PI Industries Ltd.	1.5 ml/litre	Thrips, Mealy bugs & Jassids
	Bipimain	Makhteshim – Agan India Pvt. Ltd.	1.5 ml/litre	Thrips, Mealy bugs & Jassids
	Applaud	Rallis India Ltd.	1.5 ml/litre	Thrips, Mealy bugs & Jassids
	Benz	Nagarjuna Agrichem Ltd.	1.5 ml/litre	Thrips, Mealy bugs & Jassids
	Blaze	Indofil Industries Ltd.	1.5 ml/litre	Thrips, Mealy bugs & Jassids
Carbofuran 3G	Fury	Nagarjuna Agrichem Ltd.	12.5 kg /hectare	Soil borne pests
	Furadan 3G	FMC India Ltd.	12.5 kg /hectare	Soil borne pests
	Tata Furan	Rallis India Ltd.	12.5 kg /hectare	Soil borne pests
	Difuran 3G	PI Industries Ltd.	12.5 kg /hectare	Soil borne pests
	Buroan	Sree Ramcides Chemicals Pvt. Ltd.	12.5 kg /hectare	Soil borne pests
Cartap hydrochloride 50% SP	Nidan	Crystal Phosphates Ltd.	1 g/litre	Sucking pests
	Kritap	Krishi Rasayan	1 g/litre	Sucking pests
	Caldan	Dhanuka Agritech Ltd.	1 g/litre	Sucking pests
	Cartriz	Biostadt India Ltd.	1 g/litre	Sucking pests
	Sanvex	Nagarjuna Agrichem Ltd.	1 g/litre	Sucking pests
	Mikata	Chemtura Chemicals India Pvt. Ltd.	1 g/litre	Sucking pests
Chlorfluazuron 5.4% EC	Atabron	United Phosphorus Ltd.	1.5 ml/litre	Caterpillars
Chlorantaniliprole 18.5 W/W SC	Coragen	E.I.Dupont India Pvt. Ltd.	0.3 ml/litre	Caterpillars
Chlorpyriphos 15% + Alphamethrin 1 % EC	Zero	Crystal Phosphates Ltd.	1.5 ml /litre	Jassids, Termites, Caterpillars
	Twins	Sulphur Mills	1.5 ml/litre	Jassids, Termites, Caterpillars
Chlorpyriphos 20 % EC	Dhanvan	Dhanuka Agritech Ltd.	3-4 ml /litre	Termites / Cockchafer

Contd....

	Trade Name	Company	Dose	Target Pest
	Tafaban	Rallis India Ltd.	3-4 ml /litre	Termites / Cockchafer
	Preman	Makhteshim – Agan India Pvt. Ltd.	3-4 ml /litre	Termites / Cockchafer
	Mukka	Chemtura Chemicals India Pvt. Ltd.	3-4 ml /litre	Termites / Cockchafer
	Force	Nagarjuna Agrichem Ltd.	3-4 ml /litre	Termites / Cockchafer
	Chlorovip	Godrej Agrovet Ltd.	3-4 ml /litre	Termites / Cockchafer
	Classic	Cheminova India Ltd.	3-4 ml /litre	Termites / Cockchafer
	Robon	SDS Ramcides Crop Science Pvt. Ltd.	3-4 ml /litre	Termites / Cockchafer
	Hilban	Hindustan Insecticides Ltd.	3-4 ml /litre	Termites / Cockchafer
	Kissan	Krishi Rasayan	3-4 ml /litre	Termites / Cockchafer
	Tricel	Excel Crop Care Ltd.	3-4 ml /litre	Termites / Cockchafer
	Dursban	Dow Agro Sciences India Pvt. Ltd.	3-4 ml /litre	Termites / Cockchafer
Chlorpyriphos 50 % EC + Cypermethrin 5 % EC	Satisfy	Godrej Agrovet Ltd.	1 ml/litre	Jassids, Termites, Looper Caterpillars
	Preman Super	Makhteshim – Agan India Pvt. Ltd.	1 ml/litre	Jassids, Termites, Looper Caterpillars
	Nurocombi	Cheminova India Ltd.	1 ml/litre	Jassids, Termites, Looper Caterpillars
	Gambhir	Sree Ramcides Chemicals Pvt. Ltd.	1.25 ml/litre	Jassids, Termites, Looper Caterpillars
	Cannon	Nagarjuna Agrichem Ltd.	1 ml/litre	Jassids, Termites, Looper Caterpillars
	Ulka	Biostadt India Ltd.	1 ml/litre	Jassids, Termites, Looper Caterpillars
	Cyklon	GSP Crop Science Pvt. Ltd.	1 ml/litre	Jassids, Termites, Looper Caterpillars
	Hilhunter	Hindustan Insecticides Ltd.	1 ml/litre	Jassids, Termites, Looper Caterpillars
	Hamla	Gharda Chemicals Ltd.	1 ml/litre	Jassids, Termites, Looper Caterpillars
	Nurani – 505	Anu Products	1 ml/litre	Jassids, Termites, Looper Caterpillars
	Koranda	Rallis India Ltd.	1 ml/litre	Jassids, Termites, Looper Caterpillars
	Anth	Krishi Rasayan	1 ml/litre	Jassids, Termites, Looper Caterpillars

Contd....

	Trade Name	Company	Dose	Target Pest
	Mega – 505	Chemtura Chemicals India Pvt. Ltd.	1 ml/litre	Jassids, Termites, Looper Caterpillars
	Terror	Isagro(Asia) Agrochemicals Pvt. Ltd.	1 ml/litre	Jassids, Termites, Looper Caterpillars
	Nurrelle – D	Dow Agrosciences India Pvt. Ltd.	1 ml/litre	Jassids, Termites, Looper Caterpillars
Chlothianidin 50% WDG	Dantotsu	Sumitomo Chemical India Ltd.	0.15 g /litre	Sucking pests
Cypermethrin 10 % EC	Record	Sree Ramcides Chemicals Pvt. Ltd.	0.75 ml/litre	Sucking and leaf eating pests
	Jackpot-10	Crystal Phosphates Ltd.	0.75 ml/litre	Sucking and leaf eating pests
	Hilcyprin	Hindustan Insecticides Ltd.	0.75 ml/litre	Sucking and leaf eating pests
	Super Mastan	Super Agro India	0.75 ml/litre	Sucking and leaf eating pests
	Shakti - 10	Cheminova India Ltd.	0.75 ml/litre	Sucking and leaf eating pests
	Cypercel - 10	Excel Crop Care Ltd.	0.75 ml/litre	Sucking and leaf eating pests
	Starcip-10	Swal Corporation Ltd.	0.75 ml/litre	Sucking and leaf eating pests
	Ustaad	United Phosphorus Ltd.	0.75 ml/litre	Sucking and leaf eating pests
	Nagcyper-10	Karnataka Agrochemicals	0.75 ml/litre	Sucking and leaf eating pests
	Supervip-10	Godrej Agrovet Ltd.	0.75 ml/litre	Sucking and leaf eating pests
	Ripcord	BASF India Ltd.	0.75 ml/litre	Sucking and leaf eating pests
	Super killer -10	Dhanuka Agritech Ltd.	0.75 ml/litre	Sucking and leaf eating pests
Cypermethrin 25% EC	Mortal-25	Chemtura Chemicals India Pvt. Ltd.	0.5 ml/litre	Sucking and leaf eating pests
	Hilcyprin 25	Hindustan Insecticides Ltd.	0.5 ml/litre	Sucking and leaf eating pests
	Cymbush	Syngenta India Ltd.	0.5 ml/litre	Sucking and leaf eating pests
	Cypervip-25	Godrej Agrovet Ltd.	0.5 ml/litre	Sucking and leaf eating pests
	Super killer -25	Dhanuka Agritech Ltd.	0.5 ml/litre	Sucking and leaf eating pests
	Colt	PI Industries Ltd.	0.5 ml/litre	Sucking and leaf eating pests
	Kobra	GSP Crop Science Pvt. Ltd.	0.5 ml/litre	Sucking and leaf eating pests
	Shakti - 25	Cheminova India Ltd.	0.5 ml/litre	Sucking and leaf eating pests

Contd....

	Trade Name	Company	Dose	Target Pest
	Basathrin	BASF India Ltd.	0.5 ml/litre	Sucking and leaf eating pests
	Kripcord-25	Krishi Rasayan	0.5 ml/litre	Sucking and leaf eating pests
	Ustaad Super	United Phosphorus Ltd.	0.5 ml/litre	Sucking and leaf eating pests
	Trofy	Isagro(Asia) Agrochemicals Pvt. Ltd.	0.5 ml/litre	Sucking and leaf eating pests
	Ranjer	SDS Ramcides Crop Science Pvt. Ltd.	0.5 ml/litre	Sucking and leaf eating pests
	Tata Cyper	Rallis India Ltd.	0.5 ml/litre	Sucking and leaf eating pests
Deltamethrin – 11 % EC	Decis 100	Bayer Crop Science Ltd.	0.4 ml/litre	Sucking and leaf eating pests
Deltamethrin 2.8 % EC	Deuce	Isagro(Asia) Agrochemicals Pvt. Ltd.	0.5 ml/litre	Sucking and leaf eating pests
	Decis	Bayer Crop Science Ltd	0.5 ml/litre	Sucking and leaf eating pests
Dichlorvos 76%	Dash	Nagarjuna Agrichem Ltd.	1 ml/litre	Mainly sucking pests
EC (DDVP)(Fumigant)				
	Divap-100	P I Industries Ltd.	1 ml/litre	Mainly sucking pests
	Doom	United Phosphorus Ltd.	1 ml/litre	Mainly sucking pests
	Coach	GSP Crop Science Pvt. Ltd.	1 ml/litre	Mainly sucking pests
	Hilvos	Hindustan Insecticides Ltd.	1 ml/litre	Mainly sucking pests
Diafenthiuron 50% WP	Pegasus	Syngenta India Ltd.	0.625 ml/litre	Mites and sucking insect
Dichlorvos 76% EC (DDVP)	Luvan	Cheminova India Ltd.	1 ml/litre	Sucking pests
	Dash	Nagarjuna Agrichem Ltd.	1 ml/litre	Sucking pests
	Divap – 100	P I Industries Ltd	1 ml/litre	Sucking pests
	Doom	United Phosphorus Ltd.	1 ml/litre	Sucking pests
	Starchlor	Swal Corporation Ltd.	1 ml/litre	Sucking pests
	Coach	GSP Crop Science Pvt. Ltd.	1 ml/litre	Sucking pests
	Nuvan	Syngenta India Ltd.	1 ml/litre	Sucking pests
	Hilvos	Hindustan Insecticides Ltd.	1 ml/litre	Sucking pests

Contd....

	Trade Name	Company	Dose	Target Pest
Diflubenzuron 25% WP	Dimilin	Dhanuka Agrictech Ltd.	1g/litre	Caterpillars
	Adept	Chemtura Chemicals India Pvt. Ltd.	1g/litre	Caterpillars
Dimethoate 30% EC	Tara-909	Swal CorporationLtd.	2.5 ml/litre	Sucking pests
	Rogor	Isagro(Asia) Agrochemicals Pvt. Ltd.	2.5 ml/litre	Sucking pests
	Tafgor	Rallis India Ltd.	2.5 ml/litre	Sucking pests
	Rogor	Cheminova India Ltd.	2.5 ml/litre	Sucking pests
	Rogosri	Crystal Phosphates Ltd.	2.5 ml/litre	Sucking pests
	Robgor	Sree Ramcides Chemicals Pvt. Ltd.	2.5 ml/litre	Sucking pests
Emamectin benzoate 5% SG	Missile	Crystal Phosphates Ltd.	0.4 g/litre	Caterpillars
	Proclaim	Syngenta India Ltd.	0.4 g/litre	Caterpillars
	Safari	GSP Crop Science Pvt. Ltd.	0.4 g/litre	Caterpillars
	Pocket	Isagro(Asia) Agrochemicals Pvt. Ltd.	0.4 g/litre	Caterpillars
	Chemdoot	Cheminova India Ltd.	0.4 g/litre	Caterpillars
	Robot	Excel Crop Care Ltd.	0.4 g/litre	Caterpillars
	Rilon	Rallis India Ltd.	0.4 g/litre	Caterpillars
	Volax	Indofil Industries Ltd.	0.4 g/litre	Caterpillars
	EM 1	Dhanuka Agrictech Ltd.	0.4 g/litre	Caterpillars
	Gallup	SDS Ramcides Crop Science Pvt. Ltd.	0.4 g/litre	caterpillars
Emamectin benzoate 1.9% EC	Billo	Crystal Phosphates Ltd.	1 ml/litre	Caterpillars
Endosulfan 35 % EC	Endosri	Crystal Phosphates Ltd.	2.5 ml/litre	Sucking pests
	Thiogan	Makhteshim – Agan India Pvt. Ltd.	2.5 ml/litre	Sucking pests
	Superkill	Super Agro India	2.5 ml/litre	Sucking pests

Contd....

	Trade Name	Company	Dose	Target Pest
	Mentor	Chemtura Chemicals India Pvt. Ltd.	2.5 ml/litre	Sucking pests
	Hildan	Hindustan Insecticides Ltd.	2.5 ml/litre	Sucking pests
	Endotaf	Rallis India Ltd.	2.5 ml/litre	Sucking pests
	Thionex	Ankar Industries Pvt. Ltd.	2.5 ml/litre	Sucking pests
	Speed	Nagarjuna Agrichem Ltd.	2.5 ml/litre	Sucking pests
	Endodhan	Dhanuka Agritech Ltd.	2.5 ml/litre	Sucking pests
	Lusulfan	Cheminova India Ltd.	2.5 ml/litre	Sucking pests
	Endocel	Excel Crop Care Ltd.	2.5 ml/litre	Sucking pests
	Thiodan	Bayer Crop Science Ltd.	2.5 ml/litre	Sucking pests
			2.5 ml/litre	Sucking pests
Endosulfan 35% EC	Endohyper	Excel Crop Care Ltd.	2.5 ml/litre	Sucking pests
+ Cypermethrin 5 % EC				
	Hildan Plus	Hindustan Insecticides Ltd.	2.5 ml/litre	Sucking pests
			2.5 ml/litre	Sucking pests
Ethion 50 % EC	Srimite	Crystal Phosphates Ltd.	2.5 ml/litre	Sucking pests
	Mitvip	Godrej Agrovet Ltd.	2.5 ml/litre	Sucking pests
	Ethion	Ankar Industries Pvt. Ltd.	2.5 ml/litre	Sucking pests
	Fosmite	PI Industries Ltd.	2.5 ml/litre	Sucking pests
	Tafethion	Rallis India Ltd.	2.5 ml/litre	Sucking pests
	Challenge	Nagarjuna Agrichem Ltd.	2.5 ml/litre	Sucking pests
	Mit-500	Swal Corporation Ltd.	2.5 ml/litre	Sucking pests
			2.5 ml/litre	Sucking pests
Ethion 40 % EC	Mitplus	Swal Corporation Ltd.	2.5 ml/litre	Sucking pests
+ Cypermethrin 5 % EC				

Contd....

	Trade Name	Company	Dose	Target Pest
	Kranti	Biostadt India Ltd.	2.5 ml/litre	Sucking pests
	Colfos 405 EC	PI Industries Ltd.	2.5 ml/litre	Sucking pests
	Nagata	Rallis India Ltd.	2.5 ml/litre	Sucking pests
	Fender	PI Industries Ltd.	2.5 ml/litre	Sucking pests
			2.5 ml/litre	Sucking pests
Ethofenprox 10% EC	Punkaso	Rallis India Ltd.	2.5 ml/litre	Sucking pests
	Nukil	Dhanuka Agritech Ltd.	2.5 ml/litre	Sucking pests
Fenvelarate 20 % EC	Fenkill	United Phosphorus Ltd.	1 ml/litre	Caterpillars
	Fenval	Isagro(Asia) Agrochemicals Pvt. Ltd.	1 ml/litre	Caterpillars
	Tatafen	Rallis India Ltd.	1 ml/litre	Caterpillars
	Triumph Card – 20	Dhanuka Agritech Ltd.	1 ml/litre	Caterpillars
	Sumicidin	Sumitomo Chemical India Ltd.	1 ml/litre	Caterpillars
Fipronil 5 % SC	Regent	Bayer Crop Science Ltd.	1.5 ml/litre	Sucking pests
	Fipro Plus	SDS Ramcides Crop Science Pvt. Ltd.	1.5 ml/litre	Sucking pests
	Refree	GSP Crop Science Pvt. Ltd.	1.5 ml/litre	Sucking pests
	Stalker SC	Indofil Industries Ltd.	1.5 ml/litre	Sucking pests
	Tzed	Krishi Rasayan	1.5 ml/litre	Sucking pests
	Agadi	Makhteshim – Agan India Pvt. Ltd.	1.5 ml/litre	Sucking pests
	Fax	Dhanuka Agrictech Ltd.	1.5 ml/litre	Sucking pests
Fipronil 80% WG	Jump	Bayer Crop Science Ltd.	0.25 g/litre	Sucking pests
Flubendiamide 20% WDG	Suraksha	Nagarjuna Agrichem Ltd.	0.325 g/litre	Caterpillars
	Takumi	Rallis India Ltd.	0.325 g/litre	Caterpillars
	Fluton	PI Industries Ltd.	0.325 g/litre	Caterpillars
	Minori	SDS Ramcides Crop Science Pvt. Ltd..	0.325 g/litre	Caterpillars

Contd....

	Trade Name	Company	Dose	Target Pest
Flubendiamide 39.35 % m/m SC	Fame	Bayer Crop Science Ltd.	0.25 ml/litre	Caterpillars
	Asset	Indofil Industries Ltd.	0.25 ml/litre	Caterpillars
Imidacloprid 17.8% SL	Nagarjuna Mida	Nagarjuna Agrichem Ltd.	0.3 ml/litre	Sucking pests
	Comfidece-555	Crystal Phosphates Ltd.	0.3 ml/litre	Sucking pests
	Dharvaar	SDS Ramcides Crop Science Pvt. Ltd.	0.3ml/litre	Sucking pests
	Chemida	Cheminova India Ltd.	0.3 ml/litre	Sucking pests
	Media	Dhanuka Agrictech Ltd.	0.3 ml/litre	Sucking pests
	Jumbo	PI Industries Ltd.	0.3 ml/litre	Sucking pests
	Atom	Indofil Industries Ltd.	0.3 ml/litre	Sucking pests
	Imidacel	Excel Crop Care Ltd.	0.3 ml/litre	Sucking pests
	Tata Mida	Rallis India Ltd.	0.3 ml/litre	Sucking pests
	Imida Gold	United Phosphorus Ltd.	0.3 ml/litre	Sucking pests
	Admit	Isagro(Asia) Agrochemicals Pvt. Ltd.	0.3 ml/litre	Sucking pests
	Sensex	GSP Crop Science Pvt. Ltd.	0.3 ml/litre	Sucking pests
	Imidiate	Godrej Agrovet Ltd.	0.3 ml/litre	Sucking pests
	Imivax Vital	Chemtura Chemicals India Pvt. Ltd.	0.3 ml/litre	Sucking pests
	Hilmida	Hindustan Insecticides Ltd.	0.3 ml/litre	Sucking pests
	Josh	Krishi Rasayan	0.3 ml/litre	Sucking pests
Imidacloprid 70% WG	Royal	Rallis India Ltd.	0.15 g/litre	Sucking pests
	R Bullet	Sree Ramcides Chemicals Pvt. Ltd.	0.15 g/litre	Sucking pests
	Atom Plus	Indofil Industries Ltd.	0.15 g/litre	Sucking pests
	Pronto	Sulphur Mills	0.15 g/litre	Sucking pests

Contd....

	Trade Name	Company	Dose	Target Pest
	Admire	Bayer Crop Science Ltd.	0.15 g/litre	Sucking pests
	Pactus	Cheminova India Ltd.	0.15 g/litre	Sucking pests
Imidacloprid 30.5% SC	Chemida Plus	Cheminova India Ltd.	0.3 ml/litre	Sucking pests
Indoxacarb 14.5 %	Indoxxa	Chemtura Chemicals India Pvt. Ltd.	1 ml/litre	Caterpillars
	Dhawa	Dhanuka Agritech Ltd.	1 ml/litre	Caterpillars
	Avant	E.I.Dupont India Pvt. Ltd.	1 ml/litre	Caterpillars
	Kare	Krishi Rasayan	1 ml/litre	Caterpillars
	PI – Indoxa	PI Industries Ltd.	1 ml/litre	Caterpillars
	Lashkar	Cheminova India Ltd.	1 ml/litre	Caterpillars
	Doxagan	Makhteshim – Agan India Pvt. Ltd.	1 ml/litre	Caterpillars
Lambda Cyhalothrin 2.5% EC	Reva	Rallis India Ltd.	1 ml/litre	Sucking & Chewing Pests
	Constant	Sree Ramcides Chemicals Pvt. Ltd.	1 ml/litre	Sucking & Chewing Pests
	Judo	Crystal Phosphates Ltd.	1 ml/litre	Sucking & Chewing Pests
	Warrior	Nagarjuna Agrichem Ltd.	1 ml/litre	Sucking & Chewing Pests
Lambda Cyhalothrin 5 % EC	Instant	SDS Ramcides Crop Science Pvt. Ltd.	0.6 ml/litre	Sucking & Chewing Pests
	Karate	Syngenta India Ltd.	0.6 ml/litre	Sucking & Chewing Pests
Monocrotophos 36% SL	Monocrown	Nagarjuna Agrichem Ltd.	2.5 ml/litre	Sucking pests
	Kadett	PI Industries Ltd.	2.5 ml/litre	Sucking pests
	Sriphos	Crystal Phosphates Ltd.	2.5 ml/litre	Sucking pests
	Monostar	Swal Corporation Ltd.	2.5 ml/litre	Sucking pests
	Phoskil	United Phosphorus Ltd.	2.5 ml/litre	Sucking pests
	Hilcron	Hindustan Insecticides Ltd.	2.5 ml/litre	Sucking pests
	Monomain	Makhteshim – Agan India Pvt. Ltd.	2.5 ml/litre	Sucking pests
	Tatamono	Rallis India Ltd.	2.5 ml/litre	Sucking pests

Contd....

	Trade Name	Company	Dose	Target Pest
	Monovip	Godrej Agrovet Ltd.	2.5 ml/litre	Sucking pests
	Monokil	Ankar Industries Pvt. Ltd.	2.5 ml/litre	Sucking pests
	R C Phos	SDS Ramcides Crop Science Pvt. Ltd.	2.5 ml/litre	Sucking pests
	Monodhan	Dhanuka Agrictech Ltd.	2.5 ml/litre	Sucking pests
	Monomar	GSP Crop Science Pvt. Ltd.	2.5 ml/litre	Sucking pests
Novaluron 10 % EC	Rimon	Indofil Industries Ltd.	1.25 ml/litre	Caterpillars
	Pedestal Supra	Makhteshim – Agan India Pvt. Ltd.	1.25 ml/litre	Caterpillars
	Nova	Rallis India Ltd.	1.25 ml/litre	Caterpillars
Oxydemeton methy	Swal Corporation Ltd.	2.5 ml/litre	Sucking pests	
l 25 % EC	Starsystox			
	Metasystox	United Phosphorus Ltd.	2.5 ml/litre	Sucking pests
Phorate 10% CG	Starphor	Swal Corporation Ltd.	12.5 kg/hectare	Soil borne pests
	Thimet	Insecticides India Ltd	12.5 kg/hectare	Soil borne pests
	Foratox	PI Industries Ltd.	12.5 kg/hectare	Soil borne pests
	Dhan-10G	Dhanuka Agrictech Ltd.	12.5 kg/hectare	Soil borne pests
	Mainphorate	Makhteshim – Agan India Pvt. Ltd.	12.5 kg/hectare	Soil borne pests
	Umet	United Phosphorus Ltd.	12.5 kg/hectare	Soil borne pests
	Kurunai	Sree Ramcides Chemicals Pvt. Ltd.	12.5 kg/hectare	Soil borne pests
Permethrin 25% EC	Permasect	Coromandel International Ltd.	1 ml/litre	Caterpillars
	Signor	Heranba Industries Ltd.	1 ml/litre	Caterpillars
Profenofos 40% EC	Direct	GSP Crop Science Pvt. Ltd.	1 ml/litre	Sucking pests
+ Cypermethrin 4 % EC				
	Kilcron Plus	Crystal Phosphates Ltd.	1 ml/litre	Sucking pests
	Profigan Plus	Makhteshim – Agan India Pvt. Ltd.	1 ml/litre	Sucking pests
	Profex Super	Nagarjuna Agrichem Ltd.	1 ml/litre	Sucking pests

Contd

	Trade Name	Company	Dose	Target Pest
	Hitcel	Excel Crop Care Ltd.	1 ml/litre	Sucking pests
	Pro-Rin	Dhanuka Agrictech Ltd.	1 ml/litre	Sucking pests
	Krifos Super	Krishi Rasayan	1 ml/litre	Sucking pests
	Cypro 44	Sree Ramcides Chemicals Pvt. Ltd.	1 ml/litre	Sucking pests
	Roket	PI Industries Ltd.	1 ml/litre	Sucking pests
Quinalphos 20% EC + Cypermethrin 3 % EC	Virat	United Phosphorus Ltd.	2 ml/litre	Caterpillars
Quinalphos 25% EC	Vazra	Cheminova India Ltd.	2.5 ml/litre	Caterpillars
	Krush	Biostadt India Ltd.	2.5 ml/litre	Caterpillars
	Dhanulux	Dhanuka Agritech Ltd.	2.5 ml/litre	Caterpillars
	Ekalux	Syngenta India Ltd.	2.5 ml/litre	Caterpillars
	Flash	Indofil Industries Ltd.	2.5 ml/litre	Caterpillars
	Hilquin	Hindustan Insecticides Ltd.	2.5 ml/litre	Caterpillars
Spinosad 45% SC	Spinosad	Dow Agrosciences India Pvt. Ltd.	0.375 ml/litre	Sucking & Chewing Pests
	Conserve	Nagarjuna Agrichem Ltd.	0.375 ml/litre	Sucking & Chewing Pests
	Taffin	Rallis India Ltd.	0.375 ml/litre	Sucking & Chewing Pests
	Charge	Cheminova India Ltd.	0.375 ml/litre	Sucking & Chewing Pests
Thiacloprid 21.7 % SC	Invest	Isagro(Asia) Agrochemicals Pvt. Ltd.	0.75 ml/litre	Sucking pests
	Alanto	Bayer Crop Science Ltd.	0.75 ml/litre	Sucking pests
	Splendour	Cheminova India Ltd.	0.75 ml/litre	Sucking pests
Thiamethoxam 25% WG	Champ	Isagro(Asia) Agrochemicals Pvt. Ltd.	0.25 g/litre	Sucking pests
	Torpid	Godrej Agrovet Ltd.	0.25 g/litre	Sucking pests
	Hilstar	Hindustan Insecticides Ltd.	0.25 g/litre	Sucking pests

Contd....

	Trade Name	Company	Dose	Target Pest
	Dxtar	Nagarjuna Agrichem Ltd.	0.25 g/litre	Sucking pests
	Act 150	SDS Ramcides Crop Science Pvt. Ltd.	0.25 g/litre	Sucking pests
	Click	Indofil Industries Ltd.	0.25 g/litre	Sucking pests
	Suckgan	Makhteshim – Agan India Pvt. Ltd.	0.25 g/litre	Sucking pests
	Anant	Rallis India Pvt.. Ltd.	0.25 g/litre	Sucking pests
	Slayer	GSP Crop Science Pvt. Ltd.	0.25 g/litre	Sucking pests
	Maxima	PI Industries Ltd.	0.25 g/litre	Sucking pests
	Actara	Syngenta India Ltd.	0.25 g/litre	Sucking pests
	Renova	United Phosphorus Ltd.	0.25 g/litre	Sucking pests
	Evident	Biostadt India Ltd,	0.25 g/litre	Sucking pests
	Krioxm	Krishi Rasayan	0.25 g/litre	Sucking pests
	Caper	Cheminova India Ltd.	0.25 g/litre	Sucking pests
Thiamethoxam 12.5% + Lambda Cyhalothrin 9.5 %	Alika	Syngenta India Ltd.	1 ml/litre	Caterpillars
Thodicarb 75 % WP	Ajax	Makhteshim Agan India Pvt. Ltd.	2 ml/litre	Caterpillars
	Spiro	Rallis India Ltd.	2 ml/litre	Caterpillars
Triazophos 36 % + Deltamethrin 1 % EC	Cheris	Cheminova India Ltd.	1 ml/litre	Sucking and Chewing pests
	Spark	Bayer Crop Science Ltd.	1 ml/litre	Sucking and Chewing pests
	Shukra	Sree Ramcides Chemicals Pvt. Ltd.	1 ml/litre	Sucking and Chewing pests
	Sarpanch	GSP Crop Science Pvt. Ltd.	1 ml/litre	Sucking and Chewing pests
Triazophos 40% EC	Hostathion	Bayer Crop Science Ltd.	1.5 ml/litre	Sucking pest
	Trifos	Cheminova India Ltd.	1.5 ml/litre	Sucking pest
	Tarzan Rite	Krishi rasayan	1.5 ml/litre	Sucking pest
	March	Chemtura Chemicals India Pvt. Ltd.	1.5 ml/litre	Sucking pest
Agriculture spray oil and insects.	Hortimin G	Aries Agrovet India Ltd.	20 ml/litre	Sucking pests. It also kills eggs of mites

Contd....

Mites

1. Red spider mites *(Oligonychus coffeae)*
2. Scarlet mite (*Brevipalpus phoenicis*)
3. Pink mite (*Acaphylla theae*)
4. Purple mite (*Calacarus carinatus*)

Red Spider Mite *(Oligonychus coffeae)*

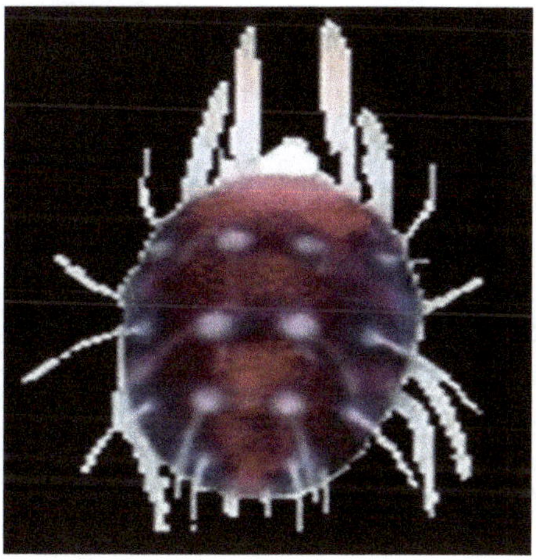

Red spider mites on the leaf surface

1. **Time of occurrence** - almost throughout the year.

2. Life-cycle

a. Eggs spherical and red, laid singly along mid rib and vein.

b. Hatch in 4-6 days.

c. Life cycle is completed in 10 – 14 days during summer.

3. Site of attack & Damage symptoms:

a. Normally attacks the upper surface of mature tea leaves. In severe cases it spreads to the undersurface and to the young leaves also.

b. Reddish spots develop on the sucking sites which later on unite to form large brown patches. Occurs mostly on the leaves upper surface.

c. The maintenance foliage turns ruddy bronze, making the infested fields distinct even from a distance.

d. In severe attack leaves may dry up and fall off.

Scarlet Mite (*Brevipalpus phoenicis*)

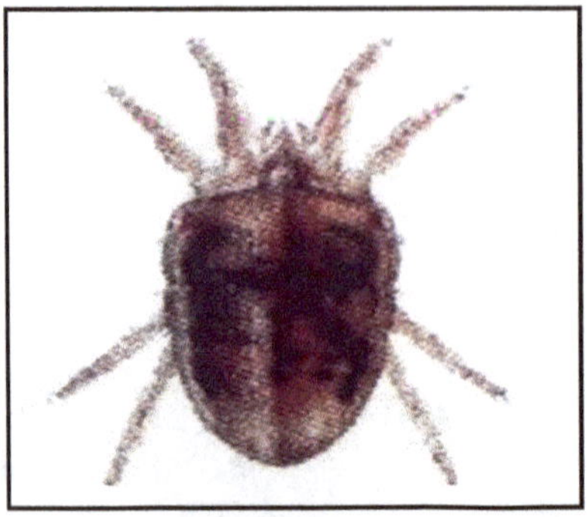

1. **Time of occurrence**: February – November
2. **Eggs and Life cycle**:
   (a) Eggs are elliptical and deep orange to purple, slim, oval.
   (b) Hatch in 11 – 13 days in March, 8 – 9 days in April and 6 – 7 days in July.
   (c) Life-cycle in 24 – 28 days in March, 21 – 25 days in April under laboratory conditions.
2. **Site of attack and Damage symptoms**:
   (a) It attacks under surface of leaves at mid rib and veins.
   (b) Yellowing of leaves and brownish discolouration of the mid rib mostly on the under surface of tea leaves.
   (c) Sometimes splitting of stem bark may occur.
   (d) Mainly young leaves are attacked.

Pink Mite (*Acaphylla theae*)

1.Time of occurrence: February – November

2.Eggs and Life-cycle:

1. Eggs are colourless or light orange, carrot shape globular with shining surface, and laid singly on the surface.
2. Incubation: 6 – 7 days in Dec. – Jan., 3 – 4 days in March, about 2 days in July – Aug.
3. Life-cycle: About 8 days in March, 6 days in July – Aug. and 13 days in Dec. – Jan. under laboratory conditions.

3. Site of attack and Damage symptoms

Attacks the under surface of tea leaves.

Tender shoots are affected, damage noticeable in advance stage.

Leaf veins and edges turn pink, affected leaves become pale, dry and leathery.

Purple Mite (*Calacarus carinatus*)

Time of occurrence: February – November

Eggs and Life-cycle

(a) Transparent, sub-hemi spherical, colourless and adult dark purple with white strips, 6 – 8 days in Jan., 4 – 5 days in March, 3 – 4 days in July – August.

(b) Nymphal Stage: 5 – 6 days in Jan., 4 – 5 days in March - April, 3 – 4 days in July – August.

(c) Life Cycle: About 7 days in July – August, 15 days in Dec. – Jan.

Site of attack & Damage symptoms

(d) Attack mostly under surface of mature tea leaves.

(e) Leaves turning purplish bronze and attaining smoky grey, dry up and fall off.

(f) Mite skins looks like dusting of white particles.

Control of Mites

Cultural Control

1. Total Control of Weeds.
2. It is necessary to prevent dust, especially in the roadside blocks, by raising boundary hedges. A row of tea bushes can be left unplucked after pruning or hedge plants like *Adhatoda vasica* or *Duranta plumeiri* or *Lantana camera* can be grown.
3. Deployment of workers should be from uninfested sections to the infested sections. Entry of cattle should also be prevented.
4. Irrigation lowers the temperature thereby reduces the chance of mite infestation.
5. Less shade prone to red spider attack, since red spider mites are thermo and phototropic.
6. Planting material: China hybrids are more susceptible than Assam hybrids.
7. Prolonged drought invites red spider mite.

Bio Pesticides

1. *Verticillium lecanii :* Liquid: 200 ml/ drum
   Powder: 800 g/ drum
1. *Paecilomyces fumosoroseus* : Liquid:- 200 ml/ drum
   Powder:- 600 g/ drum

Tea Board Approved Approved Acaricides (Plant Protection code)

Ethion 50 % EC
Dicofol 18.5 % EC
Fenazaquin 10 % EC
Fenpyroximate 5 % EC & SC
Hexythiazox 5.45 % EC
Propargite 57% EC
Spiromesifen 22.9% EC
Bifenthrin 8 % SC
Flufenzine-63% WP

NAMES OF A FEW ACARICIDES AVAILABLE IN THE MARKET (2014)

	Trade Name	Company	Dose	Target Pest
Abamectin 19% (W.W.) EC	Temprano	Chemtura Chemicals India Pvt. Ltd.	1ml/litre	Mites
	ABC	Krishi Rasayan	1ml/litre	Mites
	Abacin	Crystal Phosphates Ltd.	1ml/litre	Mites
Fenazaquin 10 % EC	Magister	E.I.Dupont India Pvt. Ltd	2 ml /litre	Mites
Bifenazate 50% EC	Benmite	Dhanuka Agritech Ltd.	0.4 g/litre	Mites
	Acramite	Chemtura Chemicals India Pvt. Ltd.	0.4 g/litre	Mites
Chlorphenapyr 10% SC	Interprid	BASF Ltd.	1.5 ml/litre	Mites
	Lepido	PI Industries Ltd.	1.5 ml/litre	Mites
Dicofol 18.5%	Colonel – S	Indofil Industries Ltd.	2.5 ml/litre	Mites
	Klin-XL	Krishi Rasayan	2.5 ml/litre	Mites
	Diumite	Ankar Industries Pvt. Ltd.	2.5 ml/litre	Mites
	Docomain	Makhteshim – Agan India Pvt. Ltd.	2.5 ml/litre	Mites
	Starkel	Swal Corporation Ltd.	2.5 ml/litre	Mites
	Hilfol	Hindustan Insecticides Ltd.	2.5 ml/litre	Mites
Diafenthiuron 50% WP	Pegasus	Syngenta India Ltd.	0.625 ml/litre	Mites and sucking insects
Fenpropathrin 30 % EC	Mighty	Excel Crop Care Ltd.	0.5 ml/litre	Mites
	Meothrin	Sumitomo Chemical India Ltd.	0.5 ml/litre	Mites
Fenpropathrin 10 % EC	Danitol	Rallis India Ltd.	1 ml /litre	Mites
Fenpyroximate 5 % EC	Neon	Rallis India Ltd	1 ml /litre	Mites
	Coronet	Cheminova India Ltd.	1 ml /litre	Mites
	Amazer	GSP Crop Science Pvt. Ltd.	1 ml /litre	Mites
	Pyromite	Excel Crop Care Ltd.	1 ml /litre	Mites

Contd....

	Trade Name	Company	Dose	Target Pest
	Mitigate	GSP Crop Science Pvt. Ltd.	1 ml /litre	Mites
	Zolomite	SDS Ramcides Crop Science Pvt. Ltd.	1 ml /litre	Mites
Fenpyroximate 5 % SC	Sedna	Rallis India Ltd	1.5 ml /litre	Mites
Hexythiazox 5.45 % W/W EC	'Dimite	Nagarjuna Agrichem Ltd.	0.8 ml/litre	Mites
	Karmit	Krishi Rasayan	0.8 ml/litre	Mites
	Macario	Isagro(Asia) Agrochemicals Pvt. Ltd.	0.8 ml/litre	Mites
	Agradoot	Cheminova India Ltd.	0.8 ml/litre	Mites
	Maiden	Biostadt India Ltd.	0.8 ml/litre	Mites
	K-Aradite	Godrej Agrovet Ltd.	0.8 ml/litre	Mites
	Endurer	Coromandel International	0.8 ml/litre	Mites
Spiromesifen 22.9% SC	Oberon	Bayer Crop Science Ltd.	1ml/litre	Mites
	Voltage	PI Industries Ltd.	1ml/litre	Mites
Agriculture spray oil	Hortimin G	Aries Agrovet India Ltd.	20 ml/litre	Sucking pests. It also kills eggs of mites and insects.
Dicofol 18.5% EC	Colonel – S	Indofil Industries Ltd.	2.5 ml/litre	Mites
	Klin-XL	Krishi Rasayan	2.5 ml/litre	Mites
	Diumite	Ankar Industries Pvt. Ltd.	2.5 ml/litre	Mites
	Docomain	Makhteshim – Agan India Pvt. Ltd.	2.5 ml/litre	Mites
	Starkel	Swal Corporation Ltd.	2.5 ml/litre	Mites
	Hilfol	Hindustan Insecticides Ltd.	2.5 ml/litre	Mites
Propargite 57% EC	Tricca	Nagarjuna Agrichem Ltd.	1.5 ml/litre	Mites
	Indomite	Indofil Industries Ltd.	1.5 ml/litre	Mites
	Simbaa	PI Industries	1.5 ml/litre	Mites

Contd....

	Trade Name	Company	Dose	Target Pest
	Acarit	Makhteshim – Agan India Pvt. Ltd.	1.5 ml/litre	Mites
	Mastamite	Chemtura Chemicals India Pvt. Ltd.	1.5 ml/litre	Mites
	Omite	Dhanuka Agritech Ltd.	1.5 ml/litre	Mites
	Xmite	SDS Ramcides Crop Science Pvt. Ltd.	1.5 ml/litre	Mites
Profenofos 50% EC	Profex	Nagarjuna Agrichem Ltd.	1 ml/litre	Mites
	Kurafen	Ankar Industries Ltd.	1 ml/litre	Mites
	Curacron	Syngenta India Ltd.	1 ml/litre	Mites
	Celcron	Excel Crop Care Ltd.	1 ml/litre	Mites
	Carina	PI Industries Ltd.	1 ml/litre	Mites
	Simcron	GSP Crop Science Pvt. Ltd.	1 ml/litre	Mites
	Auriphos	Cheminova India Ltd.	1 ml/litre	Mites
Flufenzine 63% WP	Flumite	Chemtua chemical India (P) Ltd.	1 ml/litre	Mites

Life-cycles of Few Pests

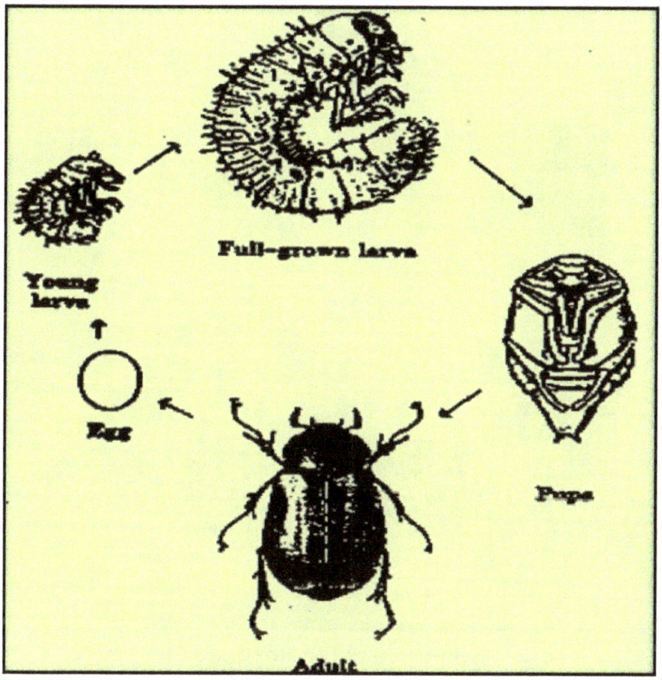

Life-cycle of Cockchafer

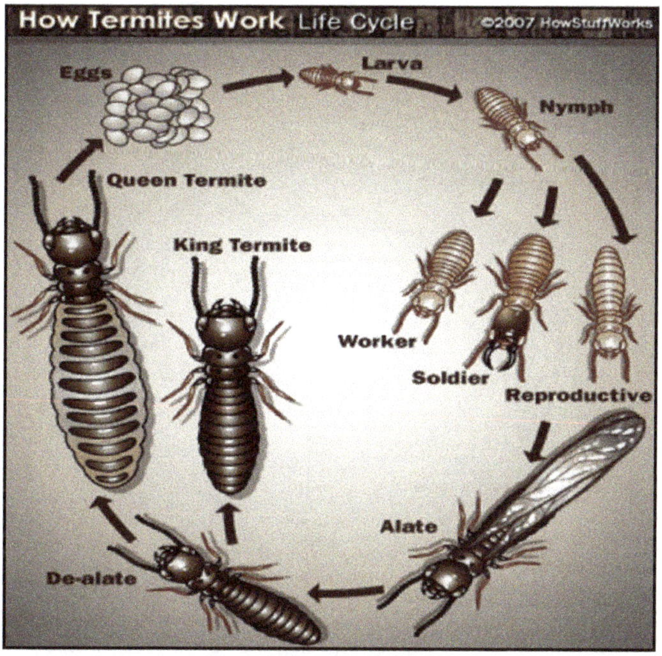

Life-cycle of Termites

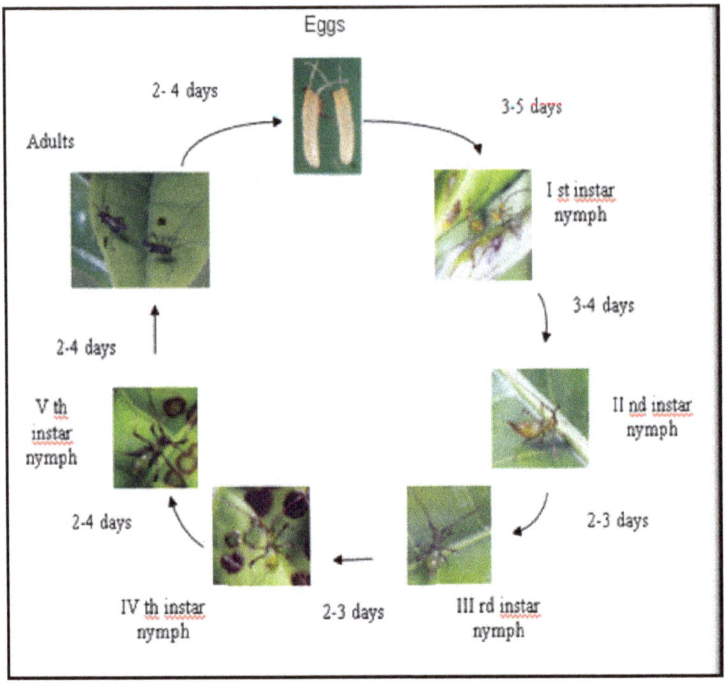

Life-cycle of Tea Mosquito Bug

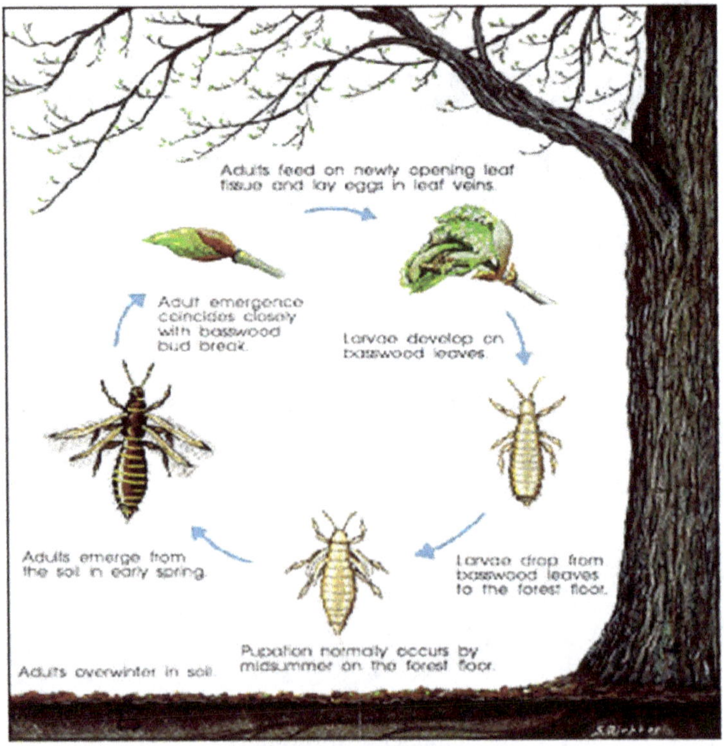

Life-cycle of Thrips

Classification of Pesticides

Based on Target Species

- **Microorganisms**

  Fungi-Fungicides

  Algae- Algaecides

  Bacteria – Batericides
- **Mites**

  Arachnids- Acaricides or Miticides
- **Insects**

  Insects – Insecticides
- **Molluscs – Molluscicides**
- **Nematodes – Nematicides**
- **Rodents – Rodenticides**

  Birds – Avicides
- **Weeds – Herbicides**

Based on Chemical Nature

Inorganic

Insecticides

Fungicides

Herbicides

Rodenticides

Organic

Insecticides

Organochlorines (cypermethrin, deltamethrin,alphamethrin, dicofol, endosulfan, lindane).

Organophosphates (dimethoate,chlorpyriphos, ethion, malathion, quinalphos,phorate) Carbamates/thiocarbamate (carbofuran, methomyl).

Herbicides

Nitrophenols (oxyfluorfen)

Chlorophenoxy (2,4-D, Simazine)

Dipyridils (thiocarbamates)

Organochlorine (dalapon)

Urea (diuron)

Other organic (paraquat)

Fungicides

Organotins (organophosphates)

Organomercurials (other organics)

CarbamatesDithiocarbamates (carbendazim, mancozeb,metham sodium, thiram, zineb, ziram)

Inorganic (copper oxychloride, copper sulphate)

Other organic (tridemorph)

Classification Based on Site of Action

Stomach Poisons

Sulphur

Contact Poisons

Quinalphos,carbamate

Systemic Poisons

Monocrotophos,phorate,dimethoate,

Phosphamidon

Fumigants

DDVP,methyl bromide,aluminium phosphide

Classification of Insecticides

(A) Based on Mode of Entry

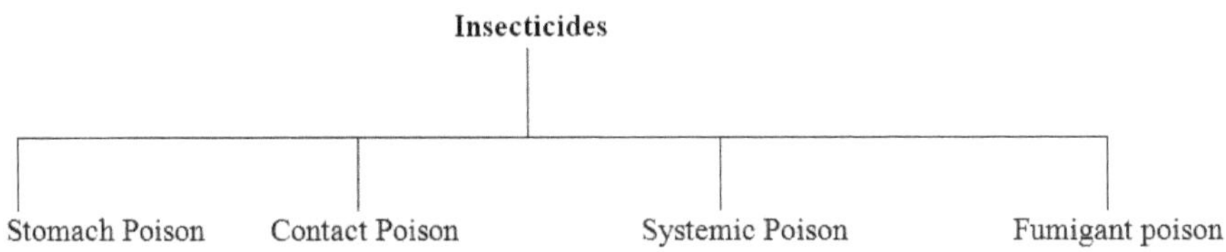

(B) Based on Mode of Action

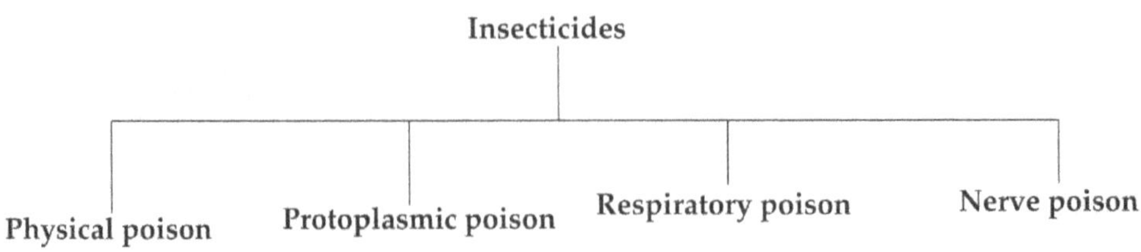

(C) Based on Formulations

- Insecticides
  - Dust
  - Granules
  - Sprays
  - Wettable powder
  - ULV
  - Aerosol
  - Fumigants
  - Poison Baits
  - EC

(D) Based on Chemical Nature

- Insecticides
  - Plant origin
    - Nicotine
    - Nornicotine
    - Pyrethrum
    - Rotemome
    - Neem
    - Sabadilla
    - Ryania
    - Acorus
    - Anabasine
  - Chemical
    - Inorganic
      - Arsenic compound
      - Fluorides
      - Sulphur
    - **Organic**
      - **(DDT,BHC,aldrin, endrin,endosulfan)**
      - **Organophosphorus Compounds (malathion, dimethoate, DDVP phorate, parathion)**
      - **Carbamates(carbofuran, aldicarb)**
      - **Di-nitrophenols (DNOC, DNOCHP)**
      - **Fumigants (methyl bromide)**

Economic Threshold Level of Few Pests

Tea mosquito bug	: 5 % infestation
Thrips	: 3 per shoot
Flush worm/Leaf roller/ Tea totrix	: 5 infested rolls / totrix per bush
Red spider mites	: 4 mites/leaf
Pink & Purple mites	: 5 mites/leaf
Blister blight	: 35 % infestation

15

TEA DISEASES

All the parts of a tea plant are infected by fungus or algae. In tea, bacterial or viral disease is not a big problem. Majority of the diseases in tea is caused by fungus. It may be mentioned that the causal organisms of diseases are also pests.

Primary Root Diseases

Name of the disease	Causal Organism
(1) Brown root Rot	*Fomes lamaoensis*
(2) Charcoal Stump Rot	*Ustulina zonata*
(3) Red Root Rot	*Poria hypobrunnea*
(4) Black Root Rot	*Rosellinia arcuata*
(5) Tarry root rot	*Hypoxylon asarcodes*
(6) Purple root rot	*Helicobasidium compactum*

Primary Stem Diseases

Name of the disease	Causal Organism
(1) Nectria	*Nectria cinnabarina*
(2) Poria	*Poria hypobrunnea*

Primary Leaf Diseases

Name of the disease	Causal Organism
(1) Black Rot	*Coticium invisum* & *Corticium theae*
(2) Blister Blight	*Exobasidium vexans*

Secondary Diseases

Secondary root diseases have pre-disposing factors for causing of the disease viz water logging, heavy drought, lightening, potash deficiency etc.

Secondary Root Diseases

Name of the disease	Causal Organism
(1) Violet root rot	Sphaerostilbe repens
(2) Thorny blight	Aglaospora sp.

Secondary Stem Diseases

Name of the disease	Causal Organism
(1) Thorny blight	Aglaospora
(2) Jew's ear fungus	Auricularia auricular

Secondary Leaf Diseases

Name of the disease	Causal Organism
(1) Brown blight	Colletotrichum camelliae
(2) Grey blight	Pestalozzia theae

Discription of a Few Primary Root Diseases

Brown Root Rot (*Fomes lamaoensis*)

Roots of affected bushes are encrusted with soil , sand and pebbles, which are held by brown coloured mycelium. The crust is difficult to remove by washing and rubbing.

Soft wood inside the brown rings decays giving honey comb like appearance.

External appearance of the tea bush

Individual bush or rarely a small group of not more than nine bushes die, leaves green to reddish in colour, withered leaves remain attached.

Symptoms at the collar region

Collar without encrustation but with mycelial growth.-

Brown soft mycelium extending up the main stem to several inches.

Symptoms on the root surface

Roots with encrustation of soil, sand, stone particles etc. The crust is difficult to remove by washing or rubbing.

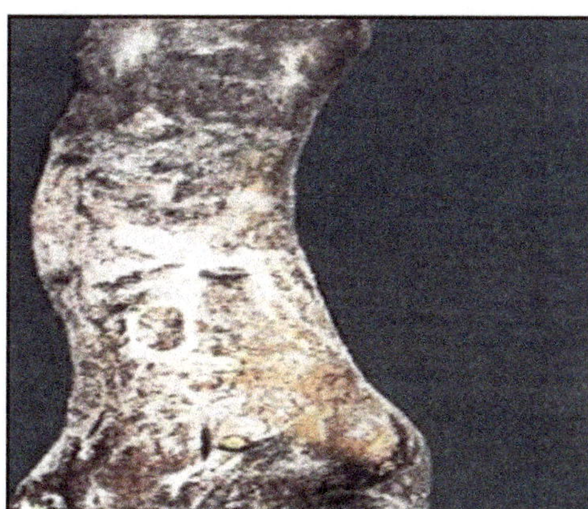

Mycelium cream to brown, general colour of root mottled brown and black.

Symptom beneath the bark

Films of brown mycelium on wood surface.

Symptom on the wood surface

Reticulations of hard brown lines, cream coloured, soft wood between them, honey comb structure; Colour of wood light yellow.

Red Root Rot (*Poria hypolateritia*)

Roots are encrusted with soil , sand and stone particles held by white mycelium. Mycelium exhibits a mottled colour combination of white , red and black. Initially the encrustation is yellow but in course of time it changes to dull grey.

Roots of the affected bushes are encrusted with soil, sand and stone particles held by white & chocolate red to black cord or sheets of mycelium. The crust can be easily removed by rubbing or washing.

External appearance of the tea bush

Individual bush or rarely a small group of not more than nine bushes die, leaves green to reddish in colour, withered leaves remain attached.

Symptoms on the root surface

Roots with encrustation of soil, sand, stone particles etc. Mycelium white to chocolate red or black cords or sheets. General colour of root mottled with red, black and white.

Symptoms beneath the bark

Thin white films of mycelium on wood surface.

Black Root Rot (*Rosellinia arcuata*)

Mycelium is like stars, white or black. Many black dots and dashes are seen underneath the bark. Fruit bodies are black in colour and spherical in shape.

External appearance of the tea bush

Individual bush or rarely a small group of not more than 9 bushes die, leaves green to reddish in colour, withered leaves remain attached.

Symptom at the collar region

Collar without encrustation but with mycelial growth. Woolly, greyish stocking of mycelium sometimes extending to the base of the branches.

Symptom on the root surface

Roots with cords or strands of mycelium on the surface. Cords cobwebby, black or greyish black, loose or adherent.

Symptom beneath the bark

Star shaped , small, white to black strands of mycelium accompanied by irregular dots and dashes.

Symptoms on the wood surface

Wood with black dots and dashes.

Special Character

1. Star like mycelium, white to black, accompanied by numerous small dots and dashes underneath the bark.
2. Black, spherical fructifications.

Tarry Root Rot (*Hypoxylon asarcodes*)

A black encrustation at the collar region like a smooth and hard black crust.. The mycelium gives a beautiful graphical representation.

External appearance of the affected bush

Individual bush or rarely a small group of not more than nine bushes die, leaves green to reddish in colour, withered leaves remain attached.

Symptoms at the collar region

Collar with encrustation (sand, stones, pebbles).Encrustation black, effused, hard surface of crust smooth and like dried black point.

Symptoms on the wood

Wood with dark bands and small reticulations of light brown thin lines.

Special Characters

1. The stems from the ground level upward to several inches develop a black encrustation. Graphs like *Aglaospora*.
2. Collar carries a smooth hard crust.

Purple Root Rot (*Helicobasidium compactum*)

Thick velvety mycelium is found on the colour region. Affected wood turns pink in colour.

External appearance of the affected bush

Individual bush or rarely a small group of not more than nine bushes die, leaves green to reddish in colour, withered leaves remain attached.

Symptoms at the collar region

Thick pad of velvety, purplish brown mycelium surrounding the collar.

Symptoms at the root zone

Cords of mycelium purplish brown, cords branched, giving a general purplish brown colour to the root.

Symptoms on the wood surface

Woods without bands, reticulations or mycelium. Colour pink to pinkish brown.

Special Character

A thick velvety mycelial pad just above the soil level often surrounding whole stem to several inches. Colour of the affected wood is pinkish.

Charcoal Stump Rot (*Ustulina zonata*)

When the bark is peeled off, white , silky mycelium is visible. Charcoal like black brittle encrustation is found on the collar . Fruit bodies develops on the collar region.

External appearance of the affected bush

Mature bush part dead and remainder apparently normal and healthy. Dead part may be completely rotten.

Symptoms at the collar region

Encrustation at first white, then greenish grey, finally black. Surface undulated, brittle, charcoal like when broken.

Symptoms at the root zone

Roots without cords or strands of mycelium on the surface. White or black cushion or lumps on the root.

Symptoms underneath the bark

Film of dull white, silky, fan like mycelium.

Description of a Few Secondary Root Diseases

Violet Root Rot (*Sphaerostilbe repens*)

This is a secondary disease and develops in the area which is waterlogged. The roots become inky black or light violet coloured. The smell of the dug root is vinegary. Leaves of the affected bushes drop off in green condition only. Fruit bodies develops at the collar region in clusters and resembles tiny flower buds.

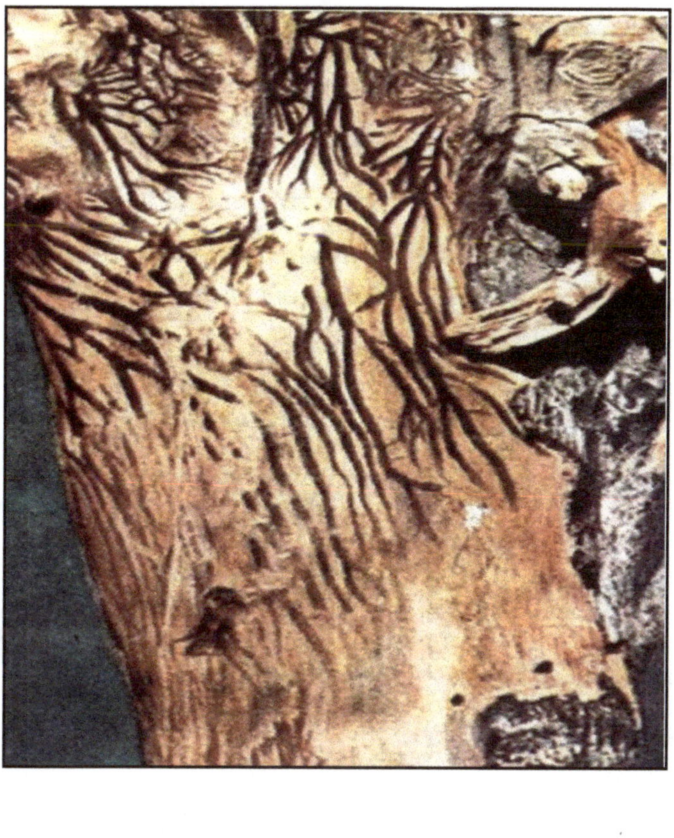

Show String Fungus (Armillaria mellea)

A split develops at the collar region upto 6 cm from the ground level. This furrows of the split is covered by white mycelium.

Roots posses brownish black rhizomorphs which looks like shoe strings.

Thorny Blight (*Aglaospora syn Tunstallia aculeata*)

After the bark is peeled off, fern like mycelium is seen. It is called thorny blight because the fruit bodies , which develop on the bark has thorn like projections.

Description of a Few Stem Diseases

Thorny Stem Blight (*Aglaospora syn Tunstallia aculeata*)

White strands of mycelium is seen on the surface underneath the bark. Black , thorn like projections are seen on the bark.

Branch Canker (*Poria hypobrunnea*)

Thin irregular brown coloured lines are seen on the wood. Infected wood becomes yellow coloured . It becomes soft and ultimately decays. Fruit bodies are yellow coloured , which turns dark grey in course of time. Fruit bodies are found at the base of the big branches.

Stem Red Rust (*Cephaleuros parasiticus*)

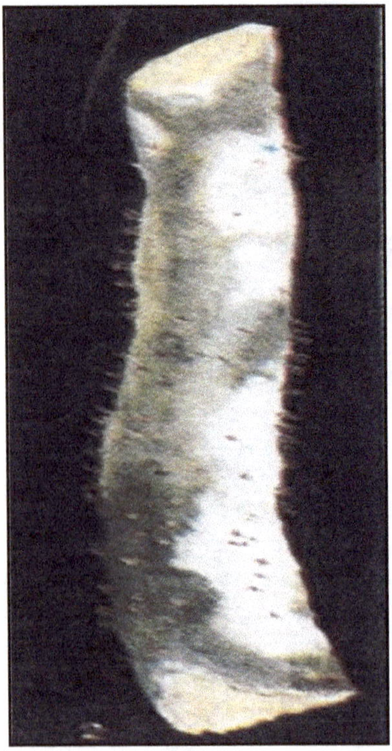

Red rust lesions are circular in shape. The fructifications are brick red or orange coloured. Sporangia are located at the apex takes the shape of a knob in which numerous spores are present.

Description of a Few Leaf Diseases

Black Rot (*Corticium invisum / Corticium theae*)

The apex and margins turns black and the leaf become sodden in the rainy season. The dry leaves develops a mottling of brown and grey colours. Two or more leaves remain attached with the help of a gummy exudation.

Leaf Red Rust (*Cephaleuros mycoides*)

Small brownish warts develop on the leaf surface which are circular in shape.

Blister Blight(*Exobasidium vexans*)

Shallow depression is found on the upper surface of the leaves , which appears like convex warts on the under surface. Greyish white spores are found inside these warts.

Brown Blight (*Colletotrichum camelliea*)

Generally old leaves are affected by this disease. On young leaves also yellowish green spots develop. Circular irregular spots develop on the leaves with grey and red mycelium.

Grey Blight

The patches on the leaves are light brown (sometimes dark brown). The shape of the patch is circular or oval. The concentric rings of mycelium are distinctly visible . On the dorsal surface of the leaves, fruit bodies (pustules) are located in concentric rings. The size of the fruit bodies are slightly bigger than the fruit bodies of Brown blight.

Thread Blight

This fungus develops white thread like strands of mycelium on the stems and under surface of the leaves.

Control of Primary Root Diseases

The control measure varies according to the condition and severity of infection.

Condition I : When only a single individual bush is infected , the infected bush should be immediately removed. Treat the soil with fungicides.

Condition II : When few bushes are infected in a small patch, remove all the affected bushes and also a row of healthy plants surrounding the patch. Treat the soil with fungicides.

Condition III : When a large area is affected, isolate the area by digging a 90 cm deep trench circling the affected area. The dug soil should be spread inside the ring. Uproot all the bushes and treat the soil with fungicides.

Control of Fungal or Algal Diseases in General

Spray fungicides on the affected bushes as per recommendation given on the fungicide list.

Tea Approved fungicides (Plant Protection Code)

Carbendazim 12% + Mancozeb 63% WP

Hexaconazole 5 % EC

Propiconazole 25 % EC

NAMES OF A FEW FUNGICIDES AVAILABLE IN THE MARKET (2014)

	Trade Name	Company	Dose	Target diseases/ target pests
Carbendazim 12% + Mancozeb 63% WP	Sixer	Dhanuka Agrotech Ltd.	2.5g/litre	Red rust, black rot etc
	Mercury	GSP Crop Science Pvt. Ltd.	2.5g/litre	Red rust, black rot etc
	Companion	Indofil Industries Ltd.	2.5g/litre	Red rust, black rot etc
	Saaf	United Phosphorus Ltd	2.5g/litre	Red rust, black rot etc
	CM-75	Krishi Rasayan	2.5g/litre	Red rust, black rot etc
	Covert	Cheminova India Ltd.	2.5g/litre	Red rust, black rot etc
	Policy	SDS Ramcides Crop Science Pvt. Ltd.	2.5g/litre	Red rust, black rot etc
Carbendazim 50% SC	Pearl	Sulphur Mills	1 ml /litre	Fungus
	Benfil	Indofil Industries Ltd.	1g/litre	Fungus
	Hilzim	Hindustan Insecticides Ltd.	1g/litre	Fungus
	Bavistin	BASF Limited	1 g /litre	Fungus
	Spot free	Chemtura Chemicals India Ltd.	1 g/litre	Fungus
	Sristone	Crystal Phosphates Ltd.	1 g/litre	Fungus
	Fungy	SDS Ramcides Crop Science Pvt. Ltd.	1 ml/litre	Fungus
	Carzim	Cheminova India Ltd.	1 g/litre	Fungus
	Arrest	Isagro (Asia) AgrochemicalsPvt. Ltd.	1 g/litre	Fungus
	Kavistin	Krishi Rasayan	1 g /litre	Fungus
	Benman	Makhteshim – Agan India Pvt. Ltd.	1 g/litre	Fungus
	Zen	Nagarjuna Agrichem Ltd.	1g/litre	Fungus
				Fungus
Carboxin 75% WP	Vitavax	Dhanuka Agrtech Ltd.	2.5 g/litre	Fungus
	Provax	Chemtura Chemicals India Ltd.	2.5 g/litre	Fungus

Contd....

	Trade Name	Company	Dose	Target diseases/ target pests
Copper Oxychloride 50% WP	Top Gun	Sulphur Mills	1.75 g/litre	Blister blight, Black rot, Red rust and other fungal diseases
	Maincop	Makhteshim – Agan India Pvt. Ltd.	2.5 g /litre	Blister blight, Black rot, Red rust and other fungal diseases
	Colombia	Super Agro India	2.5 g /litre	Blister blight, Black rot, Red rust and other fungal diseases
	Chemzole	Cheminova India Ltd.	2.5 g /litre	Blister blight, Black rot, Red rust and other fungal diseases
	Blytone	Crystal Phosphates Ltd.	2.5 g /litre	Blister blight, Black rot, Red rust and other fungal diseases
	Dhanucop	Dhanuka Agritech Ltd.	2.5 g /litre	Blister blight, Black rot, Red rust and other fungal diseases
	Blue Copper	Syngenta India Ltd.	2.5 g /litre	Blister blight, Black rot, Red rust and other fungal diseases
	Raze	Isagro (Asia) AgrochemicalsPvt. Ltd.	2.5 g /litre	Blister blight, Black rot, Red rust and other fungal diseases
	Hilcopper	Hindustan Insecticides Ltd.	2.5 g /litre	Blister blight, Black rot, Red rust and other fungal diseases
	Blitox	Rallis India Ltd.	2.5 g /litre	Blister blight, Black rot, Red rust and other fungal diseases
	Trust	Krishi Rasayan	2.5 g /litre	Blister blight, Black rot, Red rust and other fungal diseases
Copper hydroxide	Isacide	Isagro (Asia) AgrochemicalsPvt. Ltd.	0.7 g/litre	Blister blight, Black rot, Red rust and other fungal diseases
	Kocide	E.I.Dupont India Pvt. Ltd.	0.7 g/litre	Blister blight, Black rot, Red rust and other fungal diseases
	Hi-Dice	Dhanuka Agritech Ltd.	0.7 g/litre	Blister blight, Black rot, Red rust and other fungal diseases
Hexaconazole 5 % EC	Mass	Nagarjuna Agrichem Ltd.	1 ml /litre	Blister blight mainly
	Roshan	SDS Ramcides Crop Science Pvt. Ltd.	1 ml /litre	Blister blight mainly
	Alert	GSP Crop Science Pvt. Ltd.	1 ml /litre	Blister blight mainly
	Mainex – 5 EC	Makhteshim – Agan India Pvt. Ltd.	1 ml /litre	Blister blight mainly
	Malconda	Crystal Phosphates Ltd.	1 ml /litre	Blister blight mainly
	Sitara	Indofil Industries Ltd.	1 ml /litre	Blister blight mainly
	Hexzol	Excel Crop Care Ltd.	1 ml /litre	Blister blight mainly
	Krizole	Krishi Rasayan	1 ml /litre	Blister blight mainly
	Danzole	Cheminova India Ltd.	1 ml /litre	Blister blight mainly
	Contaf	Rallis India Ltd.	1 ml /litre	Blister blight mainly

Contd....

	Trade Name	Company	Dose	Target diseases/ target pests
Hexaconazole 4 % + Zineb 68%	Avtar	Indofil Industries Ltd.	1.5 g/litre	Blister blight etc
Kasugamycin 3 % SL	Ibomycin	Biostadt India Ltd.	2 ml/litre	Blister blight etc.
	Kasu-B	Dhanuka Agritech Ltd.	2 ml/litre	Blister blight etc.
Mancozeb 35% SC	Hydroman	Sulphur Mills	3 ml/litre	Fungicide
Mancozeb 75% WP	Uthane	United Phosphorus Ltd.	2.5 g/litre	Blister blight and other fungus
	Indofil M 45	Indofil Industries Ltd.	2.5 g/litre	Blister blight and other fungus
	Zeb M45	Nagarjuna Agrichem Ltd.	2.5 g/litre	Blister blight and other fungus
	Dithane –M 45	Dow Agroscience s India Pvt. Ltd.	2.5 g/litre	Blister blight and other fungus
	Bilzeb	Bayer Crop Science Ltd.	2.5 g/litre	Blister blight and other fungus
Myclobutanil 10% WP	Susthane	Dow Agroscience s India Pvt. Ltd.	0.4 g/litre	Blister blight and other fungus
Propineb 70% WP	Antracol	Bayer Crop Science Ltd.	2.5 g/litre	Broad spectrum contact fungicide
	Proximain	Makhteshim – Agan India Pvt. Ltd.	2.5 g/litre	Broad spectrum contact fungicide
Propiconazole 25 % EC	Tilt	Syngenta India Ltd.	0.5 ml/litre	Fungus
	Result	Nagarjuna Agrichem Ltd.	0.5 ml/litre	Fungus
Thiofanate Methyl 70% WP	Hilnate	Hindustan Insecticides Ltd.	1 g/litre	Powdery mildew , leaf blight etc
	Roko	Biostadt India Ltd.	1 g/litre	Powdery mildew , leaf blight etc
	Topmast	Makhteshim – Agan India Pvt. Ltd.	1 g/litre	Powdery mildew , leaf blight etc

16

WEEDS AND WEED CONTROL

The term weed is applied to " all those plants for which man has found no use and which in their evolution have developed such power of aggression, persistence or reproduction as to make them a menace to the best possible development of crop. Weed can also be defined as certain undesirable species of plants which are unwanted, unuseful often prolific and persistently interfere with agricultural operation , increase labour , add to cost and reduce crop yield.

Harmful Effect of Weeds

1. In young tea weeds can remove as high as 270 kg of Nitrogen per hectare.
2. Few species of weeds serve as alternate hosts of pests .
3. In Sri Lanka 9 per cent crop loss have been reported in tea industry.
4. In Indian tea, 6 to 12 per cent crop loss has ben reported.
5. Weeds competes with tea for nutrients, moisture and light.
6. Weed infestation can restrict branching and frame development in young tea.
7. Weed coverage on plucking surface disrupts photosynthesis and productivity.
8. Weed hindrance on the path of the pluckers reduce pluckers' productivity.
9. Reduce water flow in the drains.

Types of Weeds

Annuals:

1. Complete life cycle within one season of two to four months.
2. Produce large quantity of seeds.
3. Growing period coincides with flushing period of tea.

   e.g. Ageratum conyzoides

Biennials:

1. Complete life cycle in two years.
2. Seed settling and flowering starts from second year.
   e.g.-Cassia sophera

Perennial:

1. They are persistent type.
2. Live more than two years.
   e.g.- Axonopus compressus

For chemical control of weeds , they are categorised in two groups-

MONOCOTS (mainly grasses)

DICOTS (broad leaved weeds)

Critical Period of Weed Competition

The critical period of weed competition is tea is April to September. Utmost care is needed to be taken to control weeds during this period so that the productivity of tea is not affected.

Control Methods

1. Cultural method
2. Manual /Mechanical method
3. Mechanical method
4. Biological method
5. Chemical method

Cultural Control

Mulching

Mulch is a protective covering of material maintained on soil surface. Mulching has smothering effect on weed control by excluding light from the photosynthetic portions of a plant and thus inhibiting the top growth.

Raising Cover Crops

Mainly in young tea after land preparation green cropping or cover cropping is done.

Cover crop is generally a legumionous plant having "nif" genes, so it can fix nitrogen in soil.

It provides organic matter to the soil. Beside increasing soil fertility , it also reduces weed growth.

Manual / Mechanical Control

Hand Weeding

It is done by physical removal or pulling out of weeds by hand .

Sickling

It is also done by hand with the help of sickle to remove the top growth of weeds to prevent seed production and to starve the underground parts.

Tillage

Tillage removes weeds from the soil resulting in their death. It may weaken plants through injury of root and stem pruning, reducing their competitiveness or regenerative capacity.

Hoeing

Hoe has been the most appropriate and widely used weeding tool for centuries. It is however, still a very useful implement to obtain results effectively and cheaply.

Cheel Hoeing

Cheel hoeing or cheeling using a half moon hoe is mostly followed in young tea is not only expensive but often leads to greater infestation of weeds. Primarily it is due to exposure of the top soil layer and the weed seed present therein. Exposure of soil also leads to soil erosion.

Biological Control

Biological control is done by using living organism's *viz.*, insects, disease organisms, herbivorous fish, snails or even competitive plants.

In biological control method, it is not possible to eradicate weeds but weed population can be reduced. This method is not useful to control all types of weeds. This method is still in the experimental stage.

Chemical Control

Chemical control is done by using chemical herbicides.

Classification of Herbicides

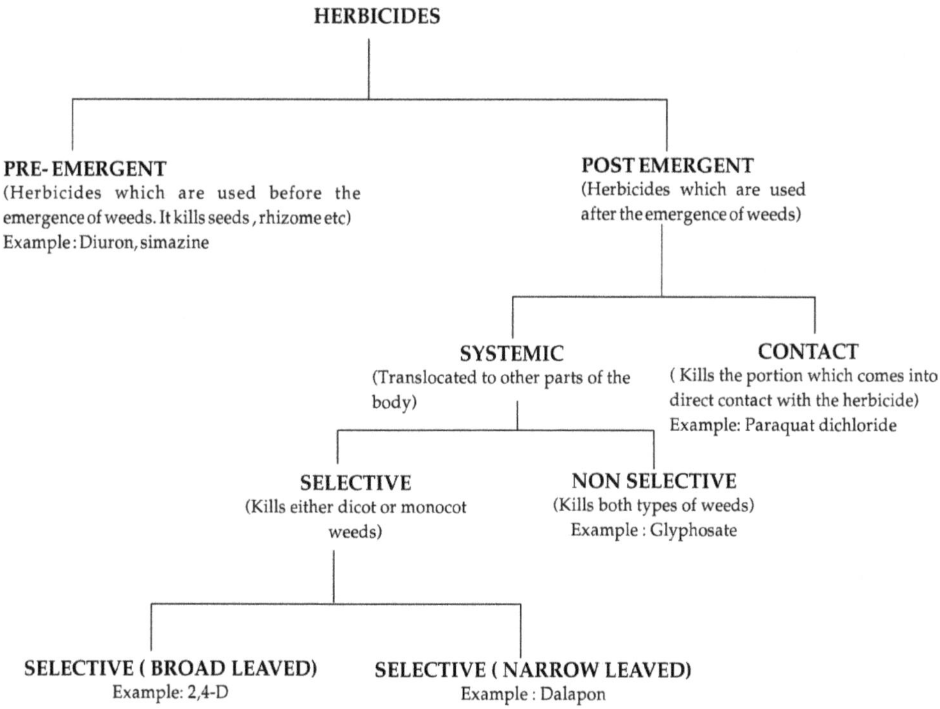

Characteristics of a Few Herbicides

Name of the Herbicide	Active ingredient	Solubility in water	Melting Point	Toxicity
2,4-D	80 %	6.2×10^2 mg/litre	140.5ºC	* * *
Dalapon	85 %	5.02×10^5 mg/litre	Decomposes At 166.5ºC	
Paraquat	24%	5×10^5 mg/litre	225-227ºC	* * *
Glyphosate	41%	1.2×10^5 mg/litre	230ºC	* *

Toxicity based on Oral LD_{50} value for male rats :

*** 51 - 500 mg / kg of body weight

** 501 to 5000 mg / kg of body weight

* › 5000 mg / kg of body weight

2,4-D = 2,4-dichlorophenoxy acetic acid (2,4-D is a hormone)

Dalapon = 2,2- dichloro propionic acid

Paraquat = 1-1′dimethyl – 4,4'- bipyridilium

Glyphosate = N- phosphonomethyl glycine (Glyphosate is an amino acid)

Mode of Action of Various Herbicides

2,4-D – Readily absorbed by roots and leaves. Moves down to phloem along with carbohydrates manufactured by leaves. 2,4_D stops cell division and induces swelling of the cells. Rate of photosynthesis is reduced. Absorption of water and soil nutrients and food transportation is disrupted.

Glyphosate – It depletes soluble sugra and starch content in the rhizome and provides complete control by checking shoot regeneration.

Paraquat – It kills the tissue by formation of H_2O_2, by oxidation process and series of chain reactions.

Dalapon – It is strong acid and protein precipitant and interferes with the synthesis of pantothenic acid , B- Vitamin used for growth.

Diuron – It interferes with Nitrogen metabolism and inhibits photosynthesis . The chemical causes loss of turgor , chlorosis and progressive die-back of leaves.

Herbicide Doses

Tea Board approved herbicides (Plant Protection code)

Glyphosate 41 % SL

Glyphosate 71 % SG

Glufosinate ammonium 13.5 % SL

Oxyfluorfen 23.5 % EC

Paraquat dichloride 24 % WSC

NAMES OF A FEW HERBICIDES AVAILABLE IN THE MARKET (2014)

	Trade Name	Company	Dose	Target Pests/ action
2,4-D Amine Salt 22.5% SL	Ankamine	Ankar Industries Ltd.	2.5 ml/litre	Broad leaved weeds
	Kay-M	Krishi Rasayan	2.5 ml/litre	Broad leaved weeds
	Kem Amine – 22.5 %	Suvochem Industries	2.5 ml/litre	Broad leaved weeds
	Weednash 22.5 %	Godrej Agrovet Ltd.	2.5 ml/litre	Broad leaved weeds
2,4-D Amine Salt 58 % SL	2,4-D Main	Makhteshim- Agan India Pvt. Ltd.	1.5 ml/litre	Broad leaved weeds
	Kay-N	Krishi Rasayan	1.5 ml/litre	Broad leaved weeds
	Kem Amine – 58%	Suvochem Industries	1.5 ml/litre	Broad leaved weeds
	Weednash – 58%	Godrej Agrovet Ltd.	1.5 ml/litre	Broad leaved weeds
2,4-Sodium Salt 80% WP	Kay-D	Krishi Rasayan	4 g/litre	Broad leaved weeds
	Fernoxone	Syngenta India Ltd.	4 g/litre	Broad leaved weeds
	Dhanuka Weedmar	Dhanuka Agritech Ltd.	4 g/litre	Broad leaved weeds
	Herboclin	Ankar Industries Ltd.	4 g/litre	Broad leaved weeds
	Kem-D	Suvochem Industries	4 g/litre	Broad leaved weeds
	Weednash	Godrej Agrovet Ltd.	4 g/litre	Broad leaved weeds
	2,4-D Agan	Makhteshim- Agan India Pvt. Ltd.	4 g/litre	Broad leaved weeds
Glyfosinate Amonium	Basta	Bayer Crop Science Ltd.	12.5 ml/litre	Pre-emergent weeds
Glyphosate 41% SL	Globus	Nagarjuna Agrochemicals Ltd.	5 ml/litre	Broad spectrum non-selective herbicide
	Sweep	United Phosphorus Ltd.	5 ml/litre	Broad spectrum non-selective herbicide
	Noweed	Dhanuka Agritech Ltd.	5 ml/litre	Broad spectrum non-selective herbicide
	PI Glypho	PI Industries Ltd.	5 ml/litre	Broad spectrum non-selective herbicide
			5 ml/litre	Broad spectrum non-selective herbicide
	Weedall	Dow Agrosciences India Pvt. Ltd.	5 ml/litre	Broad spectrum non-selective herbicide
	Krup	Krishi Rasayan	5 ml/litre	Broad spectrum non-selective herbicide

Contd....

	Trade Name	Company	Dose	Target Pests/ action
	Clinton	Crystal Phosphates Ltd.	5 ml/litre	Broad spectrum non-selective herbicide
	Brake	Biostadt India Ltd.	5 ml/litre	Broad spectrum non-selective herbicide
	Glycel	Excel Crop Care Ltd.	5 ml/litre	Broad spectrum non-selective herbicide
	Round Up	Monsanto India Ltd.	5 ml/litre	Broad spectrum non-selective herbicide
	Glyphogun	Makhteshim- Agan India Pvt. Ltd.	5 ml/litre	Broad spectrum non-selective herbicide
	Runout	G.S.P. Crop Science Pvt. Ltd.	5 ml/litre	Broad spectrum non-selective herbicide
	Clean up	Indofil Industries Ltd.	5 ml/litre	Broad spectrum non-selective herbicide
	Verve	Chemtura Chemicals India Pvt. Ltd.	5 ml/litre	Broad spectrum non-selective herbicide
	Glycides	SDS Ramcides Crop Science Pvt. Ltd.	5 ml/litre	Broad spectrum non-selective herbicide
	Glytaf	Rallis India Ltd.	5 ml/litre	Broad spectrum non-selective herbicide
	Glyphos	Cheminova India Ltd.	5 ml/litre	Broad spectrum non-selective herbicide
Glyphosate 71% SG	Globus SG	Nagarjuna Agrichem Ltd.	3 g/litre	Broad spectrum non-selective herbicide
	Excel Mera 71	Excel Crop Care Ltd.	3 g/litre	Broad spectrum non- selective herbicide
	Glyfos Dakar	Cheminova India Ltd.	3 g/litre	Broad spectrum non-selective herbicide
	Imagine	Chemtura Chemicals India Pvt. Ltd.	3 g/litre	Broad spectrum non-selective herbicide
	Glyphogun 71 SG	Makhteshim- Agan India Pvt. Ltd.	3 g/litre	Broad spectrum non-selective herbicide
	Gly Gold	SDS Ramcides Crop Science Pvt. Ltd.	3 g/litre	Broad spectrum non-selective herbicide
Paraquat dichloride 24% WSC	Chemspray	Cheminova India Ltd.	2.5 ml/litre	Contact broad spectrum herbicide
	Paranex	Makhteshim- Agan India Pvt. Ltd.	2.5 ml/litre	Contact broad spectrum herbicide
	Parawin	SDS Ramcides Crop Science Pvt. Ltd.	2.5 ml/litre	Contact broad spectrum herbicide
	Kapiq	Krishi Rasayan	2.5 ml/litre	Contact broad spectrum herbicide
	Gramoxone	Syngenta India Ltd.	2.5 ml/litre	Contact broad spectrum herbicide
	Herbucsone	Ankar Industries Ltd.	2.5 ml/litre	Contact broad spectrum herbicide
	Uniquat	United Phosphorus Ltd.	2.5 ml/litre	Contact broad spectrum herbicide
	Striker	Indofil Industries Ltd	2.5 ml/litre	Contact broad spectrum herbicide
	Firestorm	Chemtura Chemicals India Pvt. Ltd.	2.5 ml/litre	Contact broad spectrum herbicide
Quizalofop P Terfuryl 4.41% EC	Rango	Chemtura Chemicals India Pvt. Ltd.	2 ml/litre	Dicot weeds

Herbicides Cocktails

If there is variability in weed species and weed categories, herbicide cocktails are used. Few examples are given below :

Paraquat dichloride (product)- 24 % WSC 500 ml + 2,4-D (product) Amine salt – 58% SL – 300 ml per drum (1 drum = 200 litres of water).

Glyphosate (product)- 41 % SL – 800 ml + 2, 4-D (product)- 2,4-D (product) Amine salt – 58% SL – 300 ml per drum.

Glyphosate (product)- 41 % SL – 600 ml + Oxufluorfen (product)- 150 ml per drum.

Precautions to be Taken During Pesticide Spray

- Use apparels and masks.
- Storage herbicide in isolated place away from other chemicals.
- Mixing done in open place and in plastic drums.
- Keep grazing animals away from the spraying zone.
- Washing of sprayers done on non cultivated land area.

Characteristic Features of a Few Major Weeds

Abutilon indicum (L.) Sweet [Family: Malvaceae]; Local Name: *Petari, Giri phonta*

A perennial woody herb, upto 3 metres high; profusely brached.

Stem softly velvety tomentose.

Leaves alternate, simple; lamina orbicular-ovate, ± 7 cm long and 5 cm broad, entire, acute to acuminate, base cordate.

Stipules free-lateral, small.

Flowers orange or golden yellow, about 2.5 cm in diameter.

Flowers & Fruits: Rainy season and winter.

Acalypha indica L. [Family: Euphorbiaceae]; Local Name: *Muktajhuri*

Annual herbs, 25 - 75cm tall, much brached.

Leaves alternate, stipulate, simple; lamina 3 - 6 cm x 2 - 3.5 cm, generally ovate, crenate serrate, acuminate, base rounded.

Flowers in axillary spikes, green; Males are clustered towards the tip; females on lower part with large green bracts.

Flowers & Fruits: Summer and rainy season [June – October]

Achyranthes aspera L. [Family: Amaranthaceae]; Local Name: *Aapang*

Plants annual to biannual, 60 – 90 cm high.

Leaves opposite, exstipulate, simple; lamina orbicular-ovate to obovate, entire, obtuse or slightly acute, sometimes slightly notched at tip, base rounded, tomentose, velvety or glabrate.

Flowers in long slender spikes, downwardly projected, sessile, pale green, often tinged with purple; bracts ovate, acuminate, persistent, spinescent, bracteoles concave spinecent.

Utricle oblong, cylindrical, 1-seeded.

Flowers & Fruits: Throughout the year.

Ageratum conyzoides L. [Family: Compositae/ Asteraceae]; Local Name: *Oochunti.*

Annual herbs, upto 80 cm tall, odorous.

Leaves opposite in lower nodes, alternate in flowering branches, exstipulate, simple; lamina upto 12 x 9 cm, ovate, crenate-serrate, acute, rounded or cuneate at base, hairy.

Capitula in much branched corymbs, discoid. Florets all tubular, regular, white to light purple.

Cypsela (fruit) minute, conical, blackish, crowned with white hairy pappus.

Flowers & Fruits: Almost throughout the year.

Alternanthera sessilis (L.) R.Br. *ex* DC. [Family: Amaranthaceae]; Local Name: *Nunia*

Perennial herbs, prostrate to semi-erect.

Leaves opposite, exstipulate, nearly sessile; lamina 2 – 5 x 0.5 – 1.5 cm, oblong or linear oblong, entire, acute to acuminate, base cuneate.

Flowers are white to pinkish, on axillary compact short sessile spikes, sometimes head-like.

Flowers & Fruits: July to March.

Amaranthus spinosus L. [Family: Amaranthaceae]; Local Name: *Kanta Notey*

Much branched annual herbs, upto 70 cm tall with sharp axillary spines.

leaves alternate, exstipulate, simple, long petiolate; lamina 4 – 9 x 2.5 – 5 cm, ovate to ovate-lanceolate, entire, acute to shortly acuminate, base cuneate.

Flowers clustered in compact branched axillary and terminal spikes, minute, green.

Fruit with one black or brownish black seed.

Flowers & Fruits: Monsoon to post-monsoon seasons.

Argemone mexicana* L. [Family: Papaveraceae]; Local Name: *Siyalkanta

Annual herbs with yellow latex, upto 90 cm high, branched above.

Leaves alternate, exstipulate, sessile; lamina half-amplexicaul, 10 – 15 x 6 – 10 cm, lobed and dentate with yellow marginal spines.

Flowers in small terminal clusters, sessile, surrounded by leafy spiny bracts, rosaceous, golden yellow in colour.

Capsules oblong, usually prickly; seeds small spherical, black.

Flowers & Fruits: October to April.

Artemisia vulgaris* L. [Family: Compositae/ Asteraceae]; Local Name: *Naak-Nisinda, Titepatey

Suffrutescent perennial under-shrub, aromatic, pubescent hair, branches 0.5 – 2.5 m tall.

Leaves alternate, exstipulate, sessile but with stipule like lobes at the base; lamina 5 – 20 cm long, usually 3-lobed, lobes linear or linear-lanceolate, distantly serrate, acute, base cuneate to rounded, grayish green, pubescent.

Capitula globose or sub-globose, in much brached panicle, sessile or shortly pedicelate; florets brownish-yellow, all tubular.

Flowers & Fruits: August to October.

Axonopus compressus* (Sw.) P. Beauv. [Family: Gramineae/ Poaceae]; Local Name: *Chepti Ghash, Chhoto Chepti

Strictly prostrate, carpet forming perennial grass with stolons; stem with sheath compressed.

Leaves with open compressed sheathing base; blade 5 – 10 x 0.5 – 1.5 cm, linear to oblong-lanceolate, flat, entire, acute, glabrous or sparingly hairy.

Racemes erect or spreading, few; spikelets numerous, oblong or oblong-lanceolate, sessile, sub-acute or acute; grains oblong or elliptical, white.

Flowers & Fruits: August to December.

***Bidens biternata* (Lour.) Merr. & Sherff [Family: Compositae/ Asteraceae].**

Erect, annual herb, branched above, about 0.6 m tall.

Leaves opposite, 1 – 2-pinnatisect; leaflets ovate or lanceolate, serrate, acute, base cuneate.

Heads on long stout peduncles, corymbosely panicled; ray-florets sterile, white; disk-florets bisexual, yellow.

Cypsela (fruit) long, black; pappus 2 – 4 horn-like, barbed.

Flowers & Fruits: May to November.

Chromolaena odorata **(L.) R.M. King & H. Rob. [Syn.:** ***Eupatorium odoratum*** **L.; Family: Compositae/ Asteraceae]; Local Name:** ***Bon-maraa***

Suffrutescent profusely branching odorous under-shrub.

Leaves opposite, exstipulate, simple, petiolate, aromatic; lamina 6 – 9 x 2.5 – 5 cm, ovate-deltoid, deeply serrate, acuminate, base cuneate-cuspidate.

Capitula oblong in terminal corymbose panicle; florets all tubular, white to bluish-white.

Flowers & Fruits: November to March.

Commelina benghalensis **L. [Family: Commelinaceae]; Local Name: Kanshira**

Annual, small dichotomously branched diffuse herb; profusely rooting from nodes.

Leaves alternate, base sheathing, villous; lamina sessile or short petioled, 3 – 8 x 1 – 3 cm, elliptic-ovate, entire, acute to obtuse, rounded at base,

Cymes 2 – 3 flowered, enclosed in large green boat-shaped bract.

Capsule oblong, pericarp membranous, 5 seeded.

Flowers & Fruits: September to December

Croton bonplandianum **Baill. [Family: Euphorbiaceae]; Local Name:** ***Chur-churhi.***

Annual, erect, much branched, 30 – 50 cm high, woody below, with elongated lenticels and densely covered with stellate hairs, sap watery.

Leaves alternate, falsely whorled, stipulate, shortly petiolate; lamina 2 – 6 x 1.3 – 2.5 cm, ovate-lanceolate, serrate, acute, base cuneate, green.

Racemes terminal, long, often branched; female flowers few, towards the base, obscurely petiolate; males flowers on upper part, many, distinctly pedicellate; large glands on tallamus alternating with stamens.

Capsules oblong, obtuse, triquetrous, stellately hairy, break into 3-cocci.

Flowers & Fruits: Throughout the year.

Cuscuta reflexa **Roxb. [Family: Convolvulaceae]; Local Name:** ***Swarnalata***

Total stem parasite; stem pale to reddish yellow, soft, glabrous, often brown-dotted.

Leaves scaly, inconspicuous, subulate, yellow.

Flowers in small axillary racemose clusters, shortly pedicellate; corolla white; bracts oblong, obtuse.

Capsule globose.

Flowers & Fruits: January to April.

Cynodon dactylon **(L.) Pers. [Family: Poaceae]; Local Name: *Durba/ Dubba Ghaash***

Perennial mat-forming grass with innumerable anastomosing runners, rooting at nodes.

Leaves alternate with sheathing base and minute ligule; blade sessile, 2 – 3 x 0.2 – 0.3 cm, lanceolate to linear lanceolate, entire, acute to acuminate, glaucous, glabrous or hairy below.

Spikes 4, digitate on an erect peduncle; spikelets green or purplish, grains oblong free within lemma and palea.

Flowers & Fruits: December to February

Cyperus rotundus **L. [Family: Cyperaceae]; Local Name: *Mutha, Baja Mutha***

Small rosette stolaniferous herb with rhizome, perennial, root-stock creeping , 10 to 75 cm tall (including inflorescence).

Leaves arising from tuber, tristichous, bases sheathing; blade linear, tapering, acuminate, 3-nerved.

Umbels of 2-8 rayed simple or compound spikes; peduncle long erect; bracts 3, leaf like, unequal; spikelets pale yellow or brown with reddish tinge; glumes oblong, obtuse or apiculate.

Nuts trigonous, roadly obovoid, greenish black.

Flowers & Fruits: July to November

Datura metel **L. [Family: Solanaceae]; Local Name: *Dhutra***

Annual, biennial or perennial, shrubby, about 0.5 - 1 m high, stem generally green, glabrous.

Leaves alternate, exstipulate, simple, long petiolate; lamina slightly oblique-ovate to ovate-lanceolate, distantly dentate, acute, base cuneate, unequal.

Flowers axillary, solitary, shortly stout-pedicellate; calyx green, tubular; corolla funnel-shaped (infundibuliform), generally white.

Capsule globose, sometimes obscurely 2-lobed, prickly.

Flowers & Fruits: September to May

Drymaria cordata **(L.) Willd. *ex* Schult. [Family: Caryophyllaceae]; Local Name: *Aabijalo.***

Glabrous, dichotomously much branched, diffuse to trailing herb.

Leaves opposite, simple; stipules free-lateral, scaly, subulate; petiole short; lamina rounded-ovate-cordata or orbicular, entire, rounded, base cordate.

Flowers in much branched axillary and terminal dichasia; pedicel slender; calyx green; corolla inconspicuous, white, petals notched.

Capsules ovoid, trigonous, 3-valved to the base.

Flowers & Fruits: October to February.

***Eleusine indica* (L.) Gaertn. [Family: Poaceae].**

An annual grass with culms erect or geniculate, lower stem prostrate, upto 75 cm long.

Leaves crowded with sheathing bases, ligule scale-like, minute; blade linear, 10 – 30 x 0.3 – 0.5 cm, entire, acuminate, with one median nerve, glabrous or sparingly hairy.

Spikes 2 to 7, more or less digitate on a peduncle, 2.5 to 12.5 cm long.

Spikelets in two rows, sessile, ovate or ovate-oblong, green, 3- 6 flowered; grains oblong, obtusely trigonous, transversely rugose.

Flowers & Fruits: October to February.

Euphorbia hirta* L. [Family: Euphorbiaceae]; Local Name: *Pushi-didh; Bara-kerui

Small, annual, prostrate to semi-erect herb with milky latex; sometimes reddish. Much variable in shape, size and habit, sometimes upto 45 cm high; stem glandular hairy.

Leaves opposite, stipulate, almost sessile, simple; lamina 1.5 – 3.5 x 0.4 – 1.4 cm , obliquely elliptic-oblong, to lanceolate, serrate in upper part, acute, major veins meeting below tip. Flowers unisexual in cyathium with only one central female flower, arranged in axillary much-branched cymes.

Capsules hairy with reddish-brown seeds, trigonous, minutely rugose, breaks in three cocci.

Flowers & Fruits: Throughout the year

***Galinsoga parviflora* Cav. [Family: Compositae/ Asteraceae].**

Small annual herb, upto 40 cm high, much hairy, oppositely branched.

Leaves opposite, exstipulate, simple, shortly petiolate; lamina ovate, entire or few-toothed, acute, lateral nerves prominent.

Capitula or heads small, peduncled, sub-terminal; ray-florets few, dull-white; disc florets many, yellow; achenes angled or compressed.

Flowers & Fruits: Winter in plains and foot-hills; throughout the warmer period in hills.

***Helichrysum indicum* (L.) Grierson [Syn.: *Gnaphalium indicum* L.; Family: Compositae/ Asteraceae].**

Small, slender, annual herb, white woolly throughout; lower branches prostrate.

Leaves alternate, exstipulate, simple; lamina obovate to ovate-lanceolate to spathulate, entire, tip rounded, base narrowed gradually.

Capitula very small, sessile, in crowded racemes, much woolly, white, all disc.

Flowers & Fruits: December to April.

Imperata cylindrica (L.) Raeusch. [Family: Poaceae]; Local Name: *Kush, Thatch*

Perennial grass with strong underground sobols, 0.4 metre to 1.2 metres long erect to semi-erect leaves. Rootstock stoloniferous and widely creeping; culms leafy, solid, slender, thickened, pubescent or woolly at base; nodes densely bearded with erect white hair.

Leaves narrow, linear, flat, entire – very sharp, acuminate; ligules ciliate, silky at the base.

Panicle 10 cm to 20 cm long; spikelets lanceolate, brightly white-woolly.

Flowers & Fruits: March to August

Lantana camara L. [Family: Verbenaceae].

Densely much branched woody shrub with straggling branches; strongly aromatic, much prickly; young stem quadrangular.

Leaves opposite, exstipulate, petiolate; lamina 2.5 – 9 x 1 – 4 cm, ovate to oblong, sometimes slightly obovate, crenate, acute, rounded or slightly cordate at base, rugose, chartaceous.

Flowers in peduncled capitates, sessile, bracteates; outer bracts free, green, subulate; corolla tube curved; limb lobed, colour changes with age and may be orange red, scarlet, white or yellow colour.

Flowers & Fruits: Throughout the year

Leucas lavandulifolia Sm. [Syn.: *Leucas linifolia* (Roth) Spreng.; Family: Labiatae/ Lamiaceae]; Local Name: *Danda Kalash, Swet-drone*

Annual, erect, hairy non-aromatic herb, 30 – 50 cm high; stem quadrangular; internodes very long.

Leaves opposite, exstipulate, amost sessile to short petioled; lamina linear, narrowly oblong or oblong-lanceolate, entire or distantly serrate, acute, gradually narrowed at base, herbaceous

Verticellasters loose, many flowered, bracteates; mouth of calyx tube oblique; corolla distinctly bi-lipped, white; lower 2 stamens longer; carcerule of 4-nutlets, blackish-brown.

Flowers & Fruits: Throughout the year

Limnophila heterophylla (Roxb.) Benth. [Family: Scrophulariaceae].

Annual, soft aquatic herb, 15 – 45 cm in high.

Leaves opposite, exstipulate, shortly petiolate, heterophyllous; submerged leaved highly dissected to thread-like segments; arial lamina oblong, ovate, distantly serrate, acute, sparsely hairy.

Flowers axillary, solitary or few-flowered, sessile or sub-sessile; calyx 5-partite, green; corolla 2-lipped, pinkish or bluish.

Fruits hemispheric.

Flowers & Fruits: October to March.

Lindenbergia indica **Vatke [Family: Scrophulariaceae / Plantaginaceae].**

Annual, erect, oppositely branched, hispid herb.

Leaves opposite, exstipulate, simple, long petiolate; lamina ovate, crenate-serrate, obtuse to acute, glandular villous.

Flowers in small axillary clusters, shortly pedicalate, running into terminal spicate structure; calyx tubular with triangular lobes, strong-veined, green; corolla bilabiate, yellow with palate red or purple tinged.

Capsule ovoid with persistent calyx and style.

Flowers & Fruits: September to April

Melastoma malabathricum **L. [Family: Melastomataceae]; Local Name: *Dantrangi***

Bushy much-branched shrub.

Leaves opposite, exstipulate, simple, shortly petiolate; lamina 7 – 12 x 3 – 5 cm, elliptic or lanceolate or oblong-lanceolate, entire, shortly acuminate, patently strigose hairy throughout.

Flowers few in terminal clusters, almost sessile; bracts encloses the buds; corolla rosaceous, purple; stamens unequal; anthers secle-shaped, yellow with terminal dehiscence.

Fruits small, ovoid, strigose, breaking irregularly to release numerous small black seeds.

Flowers & Fruits: Throughout the year

Mikania micrantha **Kunth [Family: Compositae/ Asteraceae]; Local Name: *Goal Lata***

Shrubby much-branched climber, covers canopy of bushes; nodes swollen, oppositely branched.

Leaves opposite, exstipulate, simple, long petiolate; lamina triangular-ovate, distantly serrate, acute to acuminate, bae cuspidate-cordate to cuneate.

Capitula small, clustered in corymb-like axillary and terminal cymose panicles; involucral bracts free green; all florets tubular, dull-white. Cypsela with bristly pappus.

Flowers & Fruits: September to April.

Mimosa pudica **L. [Family: Leguminosae/ Mimosaceae]; Local Name: *Lajjabati***

A small herb with thorns. This plant is also called Touch Me Not. It is sensitive and the leaves closes if touched.

Leaves pinnately compound, Pinnae- 4 , distantly arranged, 2.5-10.0 cm long.

Petiole bristly, stipules linear- lanceolate with bristly edges . Leaflets 12-20 pairs obliquely narrow-oblong, under surface bristly.

Flowers are pink in colour situated on slender axillaries peduncles, pods jointed and prickly.

Flowers & Fruits: August to December.

Oldenlandia corymbosa* L. [Family: Rubiaceae]; Local Name: *Khet Papra

Diffuse much-branched annual herb, 20-40 cm high; stem quadrangular, slightly scabrid on angles, sparsely hairy; branches sometimes semi-erect, opposite.

Leaves opposite, simple, sessile; stipules interpetiolar; lamina linear or narrowly elliptic or elliptic-lanceolate, entire, acute, gradually narrowed to the base, slightly pubescent beneath.

Flowers 2 – 6, corymbose from the tip of a slender peduncle, distinctly pedicellate, inferior, 4-merous; calyx lobes 4, conical; corolla rotate, white.

Capsule small, ovoid, obscurely 2-lobed, crowned with persistent calyx..

Flowers & Fruits: Throughout the year.

Paspalum scrobiculatum* L. [Family: Gramineae/ Poaceae]; Local Name: *Khodoa dhan

Annual tufted grass with fibrous root system. Culms erect, decumbent at the base, 25 cm to 80 cm tall.

Leaves with sheathing bases; ligule short, membranous; blade 12.5 to 25 cm long, linear, entire, acuminate.

Panicles erect with 2 to 4 spikes, 4 to 10 cm long. Rachis much wide above; spikelets in 2 lateral rows, imbricate, orbicular or ovoid, glabrous.

Grain bi-convex.

Flowers & Fruits: April to June.

***Persicaria chinensis* (L.) H. Gross [Syn. *Polygonum chinense* L.; Family: Polygonaceae].**

Perennial undershrub, branches slender, scandant, glabrous, less branched.

Leaves alternate, simple, shortly petiolate, sometimes slightly auriculate; stipules ochreate, tubular, covering more that half of internodes, grayish brown; lamina variable in shape and size, ovate, or ovate-oblong to lanceolate, entire, acute to acuminate, base rounded, with dark markings on upper surface towards the base.

Flowers on axillary or terminal peduncled heads, sessile, bisexual; perianth light- pink or white; stamens 8 with blackish anthers.

Fruit a trigonous nutlet (achene).

Flowers & Fruits: September to March.

***Persicaria hydropiper* (L.) Delarbre [Syn.: *Polygonum hydropiper* L.; Family: Polygonaceae].**

Ascending to erect annual to bi-annual herb, 15 to 80 cm tall, stem green to reddish-brown, glabrous, looking grass-like.

Leaves alternate, simple, lower ones shortly petiolate; stipule ochreate, 0.5 – 1.5 cm long, glabrous, mouth ciliated; lamina 2 – 8 x 0.5 – 1.5 cm, narrowly lanceolate to linear, entire, acuminate, base attenuate, both surfaces reddish-glandular.

Racemes slender, interrupted, terminal and axillary; bracts small, ciliate; flowers sessile, bisexual; perianth glandular; stamens 6.

Fruit (achene) bi-convex, dark brown.

Flowers & Fruits: April to October.

Phyllanthus virgatus G. Forst. [Family: Euphorbiaceae/ Phyllanthaceae]; Local Name: _Bhui Aamla._

Erect annual, much braches herbs with leafless main shoot and leafy radiating determinate dwarf shoots, upto 60 cm tall.

Leaves alternate, simple; stipules small, scaly, free-lateral; leaves on main shoot scale-like brown, triangular-subuate, producing dwarf shoots from the axile; Leaves on dwarf shoots normal, with short petiole; lamina 0.1 – 1.5 cm long, oblong, sometimes slightly obovate, entire, short acute to obtuse-rounded, base rounded-conical, slightly oblique.

Flower-clusters axillary to normal leaves on dwarf-shoots only, mostly males, females few, pedicellate; thalamus with yellow glands; perianth scaly, yellowish; stamens monadelphous; ovary 3-lobed, styles 3, each bifurcated.

Capsule globose, smooth, obscurely 3-lobed, breaks into three cocci.

Flowers & Fruits: August to October

Scoparia dulcis L. [Family: Scrophulariaceae].

Erect annual herb to suffrutescent under-shrub, 20 to 70 cm tall; stem ridged, rigid.

Leaves opposite or whorled (generally ternate), exstipulate, simple, shortly petiolate; lamina 3 – 4 x 1.2 – 2 cm, ovate-elliptic to nearly rhomboid, serrate, acute, attenuate, glandular.

Flowers in axillary few-flowered clusters and terminally running into racemes of whorls, bisexual, nearly regular, bisexual; calyx green; corolla white.

Capsule small, ovoid or globose with persistent calyx and style.

Flowers & Fruits: July to January.

Setaria glauca (L.) P. Beauv. [Family: Gramineae/ Poaceae]; Local Name: _Pingi-natchi_

Slender loosely tufted annual grass, 3 cm to 60 cm tall, rugose, lower nodes rooting.

Leaves with sheathing bases; ligule bristly; blade flat, tapering to a slender point, glabrous or sparsely hairy, margins scabrid.

Flowers in cylindrical spikes, covered with stiff brown hairs.

Flowers & Fruits: May to November.

Sida acuta Burm. _f._ [Family: Malvaceae]; Local Name: _Berela_

Much branched erect annual herb, 40 to 70cm tall, sparsely hairy, bark rigid-fibrous with mucilaginous sap.

Leaves alternate, simple, shortly petiolate; stipules free-lateral, linear-lanceolate, unequal, entire, acute, green; lamina lanceolate to ovate-linear, rarely ovate to ovate-oblong, serrate, acute, base rounded to attenuate, 3-veined from base

Flowers solitary, axillary, ebracteate, bisexual; pedicel short on a short peduncle; epicalyx absent; calyx cup-like, 5-toothed, 5-nerved, green; petals 5, obovate, tip oblique, yellow; stamens many, monodelphous; style branches 5, stigmas capitat.

Capsule oblet-globose, with 5-ridges and persistent calyx.

Flowers & Fruits: September to January.

Solanum nigrum L. [Family: Solanaceae]; Local Name: _Kakmachi, Gurkhi_

Annual to perennial herb, variable, much branched, stem generally with one ridge, branches slender and never woody.

Leaves alternate, exstipulate, simple, petiolate; lamina 3 – 10 x 1 – 6 cm, ovate to ovate-oblong to ovate-lanceolate, entire to irregularly few serrate of obscurely lobed, acute to acuminate, base attenuate, glabrous, herbaceous.

Flowers few in extra-axillary peduncled fascicles or umbellate cymes, ebracteate; pedicels long, slender; calyx 5-lobed, green; corolla white, rotate; anthers connivent in a cone, dehiscing through terminal pore; ovary obliquely placed.

Berries globose, shining black when ripe.

Flowers & Fruits: March to November

Solanum torvum Sw. [Family: Solanaceae]; Local Name: _Got-begun_

Bushy tomentose shrub, 1.5 to 2.5 cm tall, sparsely prickly, stellate-hairy.

Leaves alternate, exstipulate, simple, long petiolate; lamina 7.5 15 x 4.5 – 12 cm, ovate or ovate-oblong, irregularly slightly sinuate, lobes and apex acute to acuminate, base slightly unequal, sparsely stellate tomentose above and densely beneath, sparsely prickly on the petiole and 1-2 prickles on the midrib beneath.

Flowers on extra-axillary, shortly peduncled, dichotomously much-branched cymes; entire inflorescence sellate-hairy, bracts absent; calyx slightly 5-toothed, green; corolla rotate, white; anthers yellow, connivent in a cone, dehisce with apical pore; ovary obliquely placed. in colour and arranged in extra-axillary dense dichotomous cymes.

Berries globose, upto 1 cm in diameter, green, turn yellow on ripening.

Flowers & Fruits: August to December.

Sonchus arvensis L. [Family: Compositae/ Asteraceae].

Annual or perennial herb with creeping stoloniferous root-stock, exudes milky latex, 30 to 100 cm tall, glandular hairy; stem hollow at internodes, ridged, umbellately branched above.

Radical leaves many in rosette, sessile,exstipulate; lamina lyrate to narrowly oblong-obovate, dentate to softly and shortly spinous; cauline leaves alternate, gradually smaller, auriculate clasping.

Capitula small in corymbose paniculate cymes, all florets ligulate, yellowish to yellow.

Flowers & Fruits: October to April.

***Spermacoce articularis* L.*f.* [Syn.: *Borreria articularis* (L.*f.*) F.N. Williams; Family: Rubiaceae].**

oppositely branched herb, procumbent to erect, upto 60 cm high; branches quadrangular, hairy.

Leaves opposite, sessile, with broad interpetiolar green stipules; lamina1.5 – 4.0 x 1 – 1.5 cm wide, elliptic-oblong, entire, acute, coriaceous, major veins parallel.

Flowers in 4 – 10 axillary clusters, epigynous; corolla rotate, white, lobes 4.

Flowers & Fruits: July to November.

***Tridax procumbens* (L.) L. [Family: Compositae / Asteraceae].**

A slender procumbent annual herb.

Leaves opposite, exstipulate, simple, petiolate; lamina ovate or ovate-lanceolate, pinnatifid, acute, hairy, fleshy.

Capitulum single on a long erect peduncle; ray florets few, marginal, light-yellow; disc florets numerous, white.

Achenes obovate or oblong, silky, brown, slightly 3- angled.

Flowers & Fruits: Throughout the year.

Urena lobata* L. [Family: Malvaceae]; Local Name:*Bun-okra

Annual, shrubby, much-branched, upto 1.5 m tall.

Leaves alternate, simple, petiolate; stipule free lateral, equal, lanceolate to obovate; lamina ovate to orbicular, 7.5 to 12.5 cm. long, lobed, stellatly pubescent beneath, midrib with a linear gland near the base.

Flowers axillary, solitary or in small clusters, very shortly pedicellate or sessile; epicalyx of 5 segments, adnate to the calyx; corolla pink with a darker centre; stamens many, monodelphous

Capsule globose, covered with stellate barbed bristles, breaks in 5 cocci.

Flowers & Fruits: July to January.

A Few Pictures of Dicot Weeds Found in Tea Gardens

Ageratum conyzoides

Alternanthera sessilis

Amaranthus spinosus

Amaranthus viridis

Artemisia dubia

Artemisia vulgaris

Bidens pilosa

Boerhavia diffusa

Borreria latifolia

Calotropis gigantea

Cassia occidentale

Chromolaena odorata

Cleome rutodosperma

Clerodendrum infortunatum

Crassocephalum crepidioides

Croton bonpladianum

Drymaria cordata

Emilia sonchifolia

Erigeron karwinskianus

Eupatorium glandulosum

Eupatorium odoratum

Galinsoga parviflora

Gnephalium indicum

Hydrocotyle Himalaica

Hyptis suaveolens

Lantana camara

Leucas indica

Melastoma malabathricum

Mikania cordata

Mimosa pudica

Mitracarpus verticillatus

Oldenlandia corymbosa

Oxalis corniculata

Peperomia pellucida

Persicaria hydropiper

Persicaria microcephala

Persicaria nepalensis

Phyllanthus urinaria

Pilea microphylla

Pouzolziz zeylanica

Scoparia dulcis

Tridax procumbens

Senna tora

A Few Pictures of Monocots Weeds Found in Tea Gardens

Axonopus compressus

Brachiaria reptans.

Cyanotis nudiflora 2

Cyanotis nudiflora

Cyperus compressus

Cyperus cyperoides

Cyperus irria

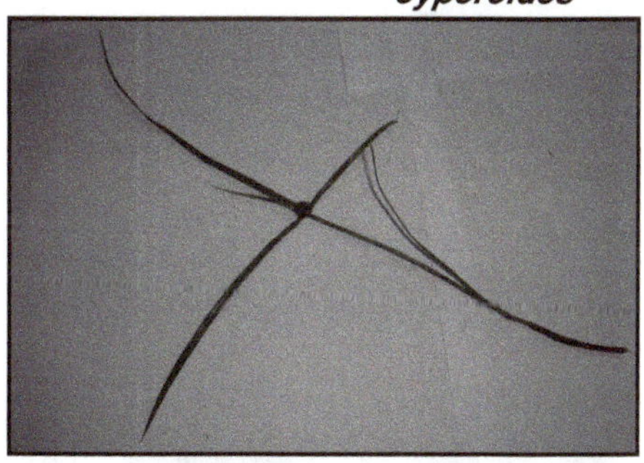

Cyperus pseudokyllinga

Cyperus sp.

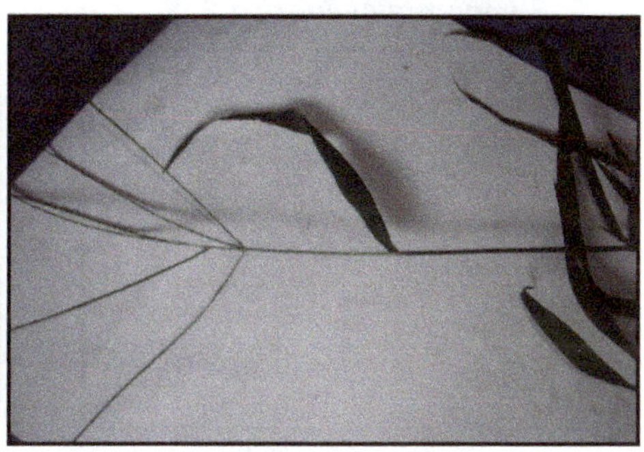

Digitaria sanguinales

Eleusine indica

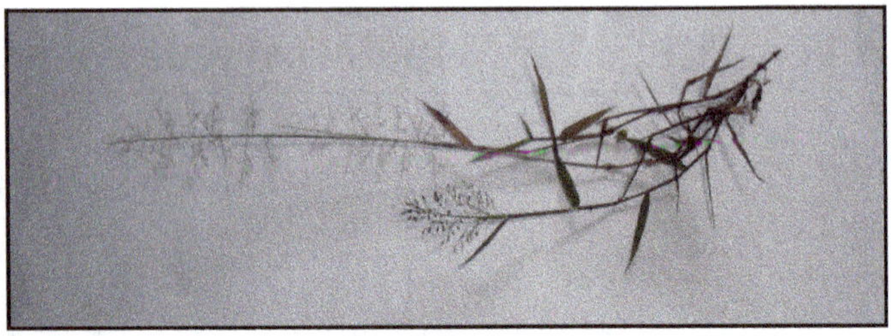

Eragrostis tenella

Eragrostis sp.

Fimbristylis miliacea

Imperata cylindrica 2

Imperata cylindrica

Kyllinga brevifolia

Murdannia nudiflora

Oplismenus burmanii

Oplismenus compositus

Panicum sp.

Paspalum sp.

Poa annua

Saccharum spontaneum

A Few Pictures of Pteridophytes Weeds Found in Tea Gardens

Adiantum sp.

Athyrium angustum

Diplazium esculentum

Dryopteris sp.

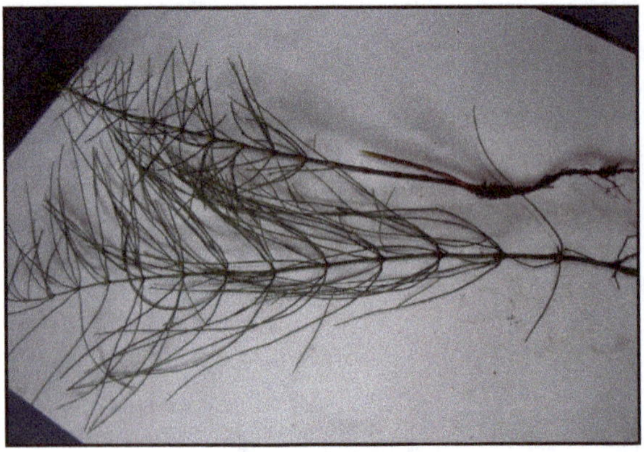

Equisetum arvense

Polypodium sp.

Pteris aquilinum

17

SPRAY TECHNIQUES

Spraying is the most popular method of pesticide and nutrient spray in a tea garden. Most pesticides are available in formulations suitable for use in sprayers.

Most common types of sprayers are as follows :

1) Knapsack sprayer
2) Compression sprayer
3) Motorised mist blower
4) Spinning disc sprayer
5) Electrostatic sprayer
6) Reciprocating power sprayer

Different Types of Nozzles

Type	Suggested use	Type of spray pattern
Flat fan	Pre-emergence & post emergence Herbicides	Fan-like pattern of medium droplets
Cone	Insecticides, acaricides and fungicides applicationin centre, fine spray	Circular, with light droplets, gives good penetration.
Flooding flat fan	Pre-emergence & post emergence	Fan-like pattern of coarse droplets herbicides

For Insecticide and Fungicide	Forherbicide
hollow cone	flat fan
duro mist	flood jet
double swril	
triple action	

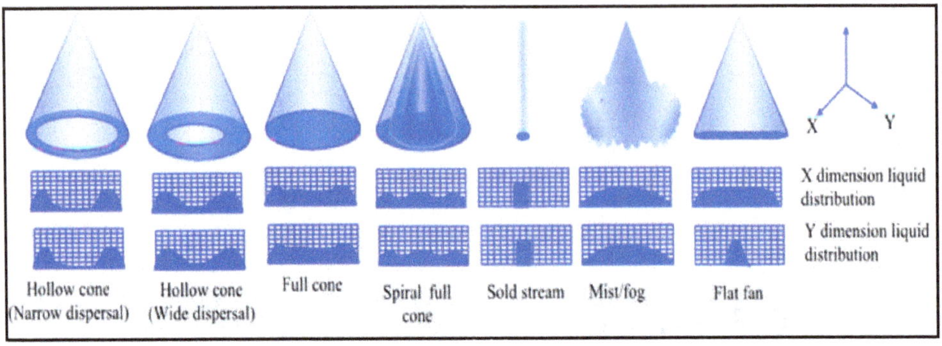

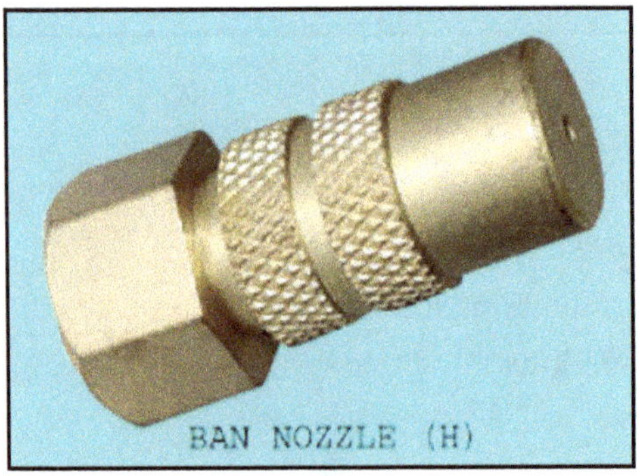

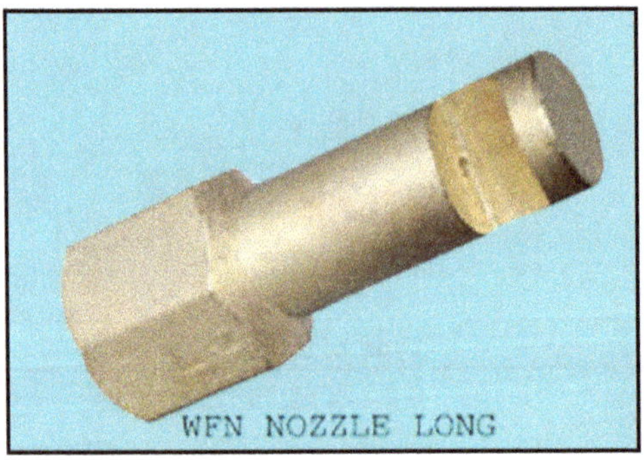

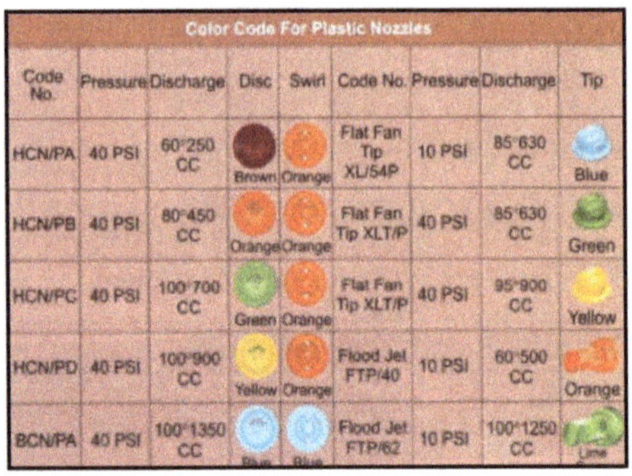

Color Code For Plastic Nozzles

Code No.	Pressure	Discharge	Disc	Swirl	Code No.	Pressure	Discharge	Tip
HCN/PA	40 PSI	60°250 CC	Brown	Orange	Flat Fan Tip XL/54P	10 PSI	85°630 CC	Blue
HCN/PB	40 PSI	80°450 CC	Orange	Orange	Flat Fan Tip XLT/P	40 PSI	85°630 CC	Green
HCN/PC	40 PSI	100°700 CC	Green	Orange	Flat Fan Tip XLT/P	40 PSI	95°900 CC	Yellow
HCN/PD	40 PSI	100°900 CC	Yellow	Orange	Flood Jet FTP/40	10 PSI	60°500 CC	Orange
BCN/PA	40 PSI	100°1350 CC	Blue	Blue	Flood Jet FTP/62	10 PSI	100°1250 CC	Lime

Plastic nozzles

Lances

Insecticide- curved or straight at 40- 60 p.s.i

Fungicide- curved or straight at 40- 60 p.s.i

Weedicide- curved or straight at 10 p.s.i

Different types of lances

Photographs of Various Sprayers

Photographs of various types of sprayers are depicted below for the benefit of the tea planters. There are many more sprayers available in the market but a few has been presented here.

Electrically operated spray machines

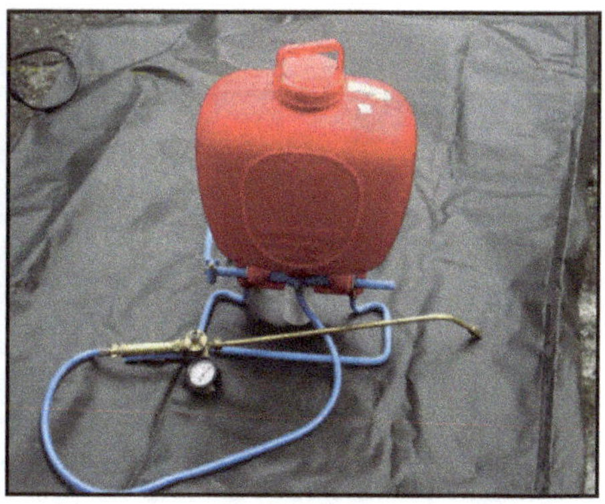

Bakpak high volume sprayer for herbicide spray- 13 litre-capacity

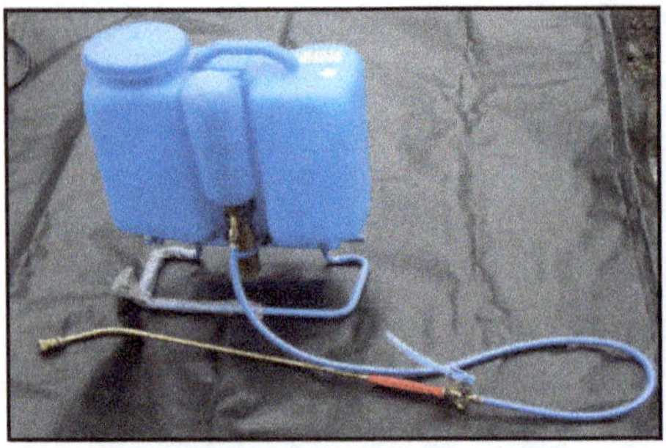

Bakpak High Tech high volume sprayer – 16 litre capacity

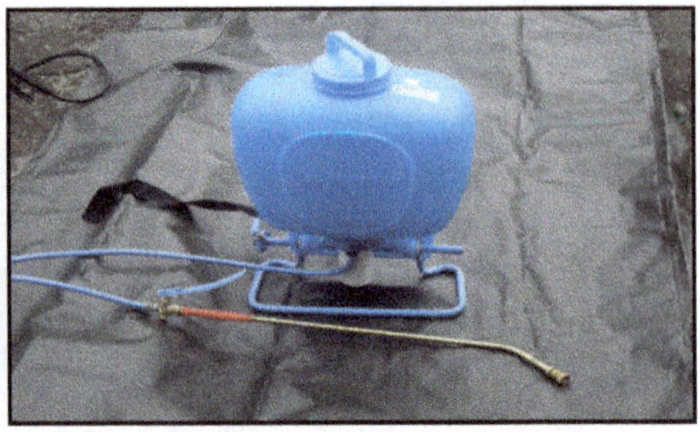

Bakpak high volume sprayer for insecticide/ fungicide / acaricide/ nutrient spray- 13 litre-capacity

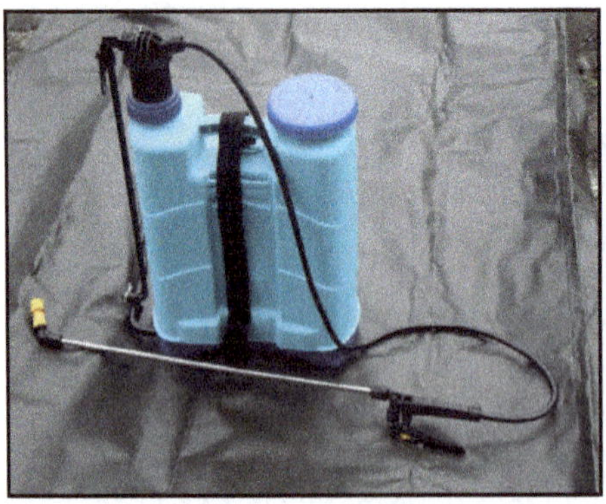

Knapsack high volume Sprayer – 16 litre- capacity

Low volume diesel operated power sprayer

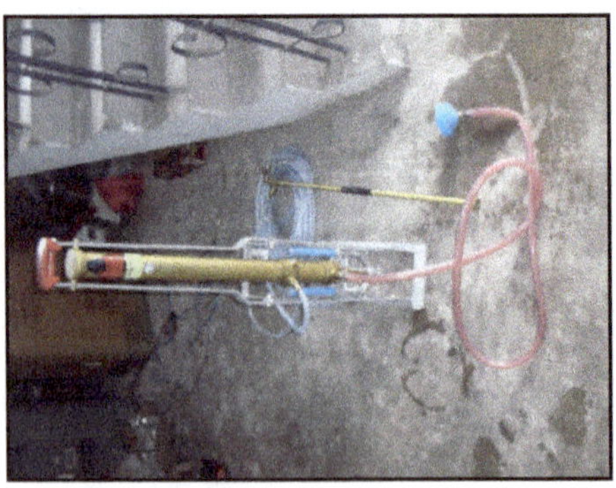

Foot sprayer for high tree spray

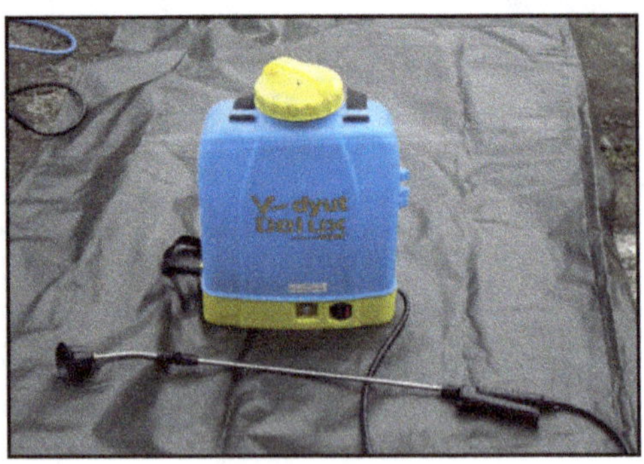

Battery operated high volume Sprayer – 16 litre- capacity

Requirement of spray fluid (lit/ ha)

Type of Tea	Hand Sprayer		Power Sprayer
	Insects	Diseases	
Pruned	200- 350	400- 600	100- 150
Skiffed	350- 500	500- 600	200- 250

Spray operation in progress

Few Tips on Spray Techniques

Faculty spray techniques can drastically reduce cost effectiveness of any product being used to improve yield potential. In many plantations insecticide sprayers are used to spray herbicides although the pressure requirement for insecticide and herbicides is quite different. Pesticides spray may be futile if the location of the target pest is not identified properly. Time of spray is also a major factor in controlling pests. Improper dilution or inappropriate discharge rate through the nozzle may upset the economy rate of spray operation.

- ❖ It is better not to spray pesticide & foliar nutrients together.
- ❖ Pesticide should be sprayed immediately after plucking but foliar nutrients & growth regulators to be sprayed 3 days after plucking.
- ❖ Use right kind of pesticide against particular pest.
- ❖ Time of spray, period of spray and method of spray should be appropriate.
- ❖ Use right kind of sprayers with right kind of nozzle and trigger.
- ❖ Never use insecticide sprayer for spray of herbicides.
- ❖ Mix each ingredient separately with water.
- ❖ Avoid spray during high sunshine hours.

Spray of Bio- Control Agents

(1) Spray when there is sufficient moisture in the phyllospheric region and soil.

(2) Spray when there is low sunshine.

(3) Wash the machines and containers before spray.

(4) No spray before and after seven days' of spray or broadcast of fungicides or bactericide.

Calibration of Sprayers

To induce economy in pest management and foliar management systems, it is essential to calibrate all the sprayers every week. Following method should be followed.

Suppose in a sprayer NMD 80450 nozzle is fitted. 80 indicates 80 degree angle and 450 indicates discharge per minute should be 450 ml. Now take a 1 litre capacity measuring cylinder. Put water in the sprayer, pump it and spray in the cylinder. Assess the discharge for 1 minute. If the discharge is more than 450 ml, the nozzle should be changed.

Herbicide Sprayers

It is not wise to spray herbicide with the help of insecticide sprayers. This main reason is there is difference in pressure. Let us study few features :

Type of sprayer		Code	Trigger Pressure
BAK PAK	SRP 23 SRP 60	TVC (pesticide)	40 psi
BAK PAK SRP 19	TCR (herbicide)	10 psi	

psi = pound per square inch

If herbicide ss sprayed with insecticide sprayers, pressure will be four time more and discharge will be just double.

Few Important Equations of Spraying

$$\text{Application rate (litres/hectare)} = \frac{Q \times 600}{B \times V}$$

where Q = Nozzle flow rate (litre/ min)

B = Swath (metres)

V = Speed (km/hr)

Spraying Capacity of Sprayers

Area covered by one sprayer per day of working is 0.4 ha for hand sprayer and 1.5 ha for power sprayer.

Requirement of Sprayers

Requirement of sprayer in estate can be calculated as per following formula:

$$\text{No. of sprayers required} = \frac{\text{Total tea area (hectare)}}{\text{Area covered by one x number of days in one round sprayer in a day}}$$

18

FEW DO' S & DON'TS IN TEA CULTIVATION

Nursery

- Nursery should not be too wet or too dry.
- Do not water nursery plants everyday , if not required.
- Use 1-2 grams of *Trichoderma* formulation and 5 grams of VAM (Vesicular Arbuscular Mycorrhiza) in each sleeve from time to time.
- Drains between the beds should at least be 120 cm deep.
- During winter, it is better to cover the beds with arched transparent polythene tent below the overhead shade.
- It is better to use netlon shade instead of thatch.

New Plantation

- Row of the tea plants should be against the slope.
- North-South or East-West is not the criterion for planting.
- Select proper clone or variety in accordance with the agro-ecological condition of the site. In low lying areas, TV 9 can be a proposition ; in high land having higher sand percentage and where water table is low, it is better to go for seed plants. In red spider prone area, Teenali 17/1/54 can be planted and TV 23 should be avoided.
- Spacing should be 90 cm X 105 cm triangular/staggered Single hedge.
- Pit size should be 45 cm deep and 30 cm dia.
- Organic matter supplements, *Trichoderma* and VAM should be used in planting pits as inputs.

- Plant height at the time of planting should be at least 45 cm.
- Following pruning schedule should be followed :

  0 Year – Training (debudding/ hand lung / decentering/partial ring barking)

  +1 Year- Recentre out (RCO)

  +2 Year – First Frame Forming Pruning at 35 cm from ground level.

  + 3 Year – Light skiffing

  +4 Year – Final Frame Forming Pruning at 45 cm from ground level
- Minimum height of plucking should be 60 cm from ground level.
- In a droughty area, *Albizzia odoratissima* and *Derris robusta* should be used as permanent shade.

Drainage

- Shallow drains are not at all helpful in a tea garden. Minimum depth of a lateral should be 105 cm, if outfall permits.
- The shape of the drains should be trapezoidal to prevent soil erosion.
- Herbicides should not be used in the drains. It may induce soil erosion.
- Lateral drains should always be against the slope.

Irrigation

- Irrigation not only increases crop potential but it helps to keep the tea bushes alive during heavy drought, especially on high land having high percentage of sand. Hence irrigation should be given during dry spell.
- Sprinkler system of irrigation is best suited in Terai and Dooars. It helps to control mites and creates favourable micro climate in the phyllospheric region.

Pruning

- In mature tea, a three year or a four year cycle or a combination of three and four year cycle should be followed. Generally young plants or the plants in a droughty area or the sections susceptible to tea mosquito bug are given a three year cycle.
- It should be ensured that least number of branches are cut during light pruning or frame forming pruning.
- Hollow center concept is dangerous. Keep maximum number of laterals as far as practicable.
- Remove thin, whippy branches, hanging branches, criss-cross branches and branches at the close proximity.
- All the cut marks should be at equal height from the ground level.
- Do not forget to use measuring stick at the time of lower pruning. Yellow painted bamboo sticks are best as measuring sticks. Measuring stick should be used in all the bushes.

- Immediately after light pruning / medium pruning / FFP , it is better to use systemic fungicides for fungal prevention.
- Bigger cut marks should be covered with bitumen paste to prevent from die back and fungal infection.
- A breather should be kept in the middle , which can be cut when sufficient sprouting is noticed in other cut branches. The breather should not be kept in very thick and naughty branches.
- It is better to bury pruning litters by digging a trench in between two rows of tea plants.
- Organic manures like cowdung/ farm yard manure and/or other organic matter supplements can be applied in between two plants by placement method. Said method is better than broadcasting of cow dung and then forking .
- If lichen and bryophytes are a problem or if the section is very much prone to thrips infestation, it is better to go for lime treatment. Make a solution with one kg of lime and 3 kg of washing soda / caustic soda in 50 litres of water and then spray after straining. Lime treatment also kills the eggs of thrips.
- The trunk of the shade trees in the light or medium pruned or FFP sections should be painted with lime upto a height of five feet. Make a solution with 10 kg of lime,500 ml of insecticide and 500 grams of copper oxychloride in 200 litres of water and paint with a jute brush.
- Medium or Deep skiff, as the requirement may be, should be inserted in each pruning cycle.

Plucking

- Addition of leaves after every six months by means of 'mother leaf plucking' in the skiffed sections should be practiced. This system is also known as 'Leaf addition'. This should be especially adopted in the sections where maintenance canopy has become scanty because of various reasons like huge pest infestation, water logging, huge drought etc.
- In the sections where sufficient maintenance foliage is available, go for 'Derh Paat ' plucking during peak season. Do not leave 'ready leaves', this can induce 'banjhiness' and reduce productivity.
- 'Plucking Tee' should be used in the pruned sections and young tea. In the skiffed sections , if the surface becomes too much uneven or if the pluckers are not skilled, it is better to use a 'Horizontal stick' so that the pluckers pluck in appropriate fashion.
- It should be ensured that the pluckers do not insert hands below the desired plucking level and do not forget to pluck the peripheral shoots.
- Plucking round should not exceed nine days in any case. It is a misconception that longer rounds improves productivity. Ideal plucking rounds per year is 32.

- Frequent weighment, weighment site nearer to the plucking site, two spring balances for each site, good quality plucking nets, clean sections, crèche facility, drinking water facilitiy, appropriate shade trees can improve pluckers' productivity.
- Flush component analysis and ballometer count should be conducted very frequently to monitor plucking rounds and schedule hormonal spray.

Soil and Nutrient Management

- Soil testing, during the pruned for each section is essential to monitor nutrient status.
- It is not possible to prepare manuring programme without analysis of soil.
- If the pH is above 5.6, application of aluminium Sulphate or pyrites would be required. If it is less than 4.5, dolomite should be applied.
- If the Organic Carbon is less than 1 % , application of FYM, oil cakes, vermicompost, bio fertilizer will help.
- If the pH is below 5.5, P_2O_5 can be applied in the form of Rock Phosphate (100 mesh). If it is above 5.5, it is better to apply as Single Super Phosphate.
- Shallow drainage increases alkalinity of soil.
- Too much application of Urea induces theanine formation in the root zone. Theanine is toxic to tea plants.
- A combination of inorganic fertilizer, organic manure and bio fertilizer is the best for proper maintenance of soil.
- The dose of Nitrogen is decided on the basis of productivity of tea of a pruning cycle and organic matter content in the soil. Potassium is related to pruning heights and available Potassium in the soil. In North East India 50 kg of P_2O_5 per hectare is good enough . Scientific manuring programme should be adopted.
- Zinc, Manganese and Boron should be applied by foliar application method. 12.5 kg of Zinc per hectare per year is quite sufficient.
- Magnesium can be applied both by foliar spray or through dolomite as ground application.
- Spray of DAP at 1 per cent concentration , 4-5 rounds a year is found to be beneficial.
- It is not necessary to spray foliar nutrients and growth regulators after every round of plucking or at least once in a month. Too much spray of nutrients can be counterproductive.
- Foliar spray should be given after two-three days of plucking.
- DAP and Zinc should not be applied together. Similarly it is better not to apply Potassium and Magnesium together.
- Overdose of fertilizers should be avoided.
- In waterlogged sections, ground application of humic & fulvic acids can be very useful.

- Sections which look exhausted, yellowish can be treated with humic acids , fulvic acids and amino acids as foliar spray.
- Ground application of fertilizers should be avoided when the soil is too dry or too moist.
- *Azospirillum, Azotobacter* and *Bacillus megatherium* , a Phosphorus solubilizing bacteria should be given at least one round for revitalizing of microbial status of soil.
- Fertilizers should not be broadcast during bright sunshine.

Growth Regulators

- To break dormancy *(banjhiness)* , spray of auxin (indole acetic acid), cytokinin, gibberellin were found to be quite effective. However, overdoing of these can be dangerous. Now-a-days many such formulations are available in the market.
- Gibberellic acid should be dissolved in acetone before tank mixing.
- To suppress huge growth of shoots, growth retardants or growth suppressants can be used.

Spray Techniques

- No spray should be done while there is bright sunshine. Early morning is the best time for spraying. However, bio control agents should be sprayed after sun down or late afternoons.
- Pesticides should be sprayed immediately after spray but foliar nutrients should be given three days after plucking.
- In the bigger capacity gardens, it is better to introduce power operated sprayers. Power sprayers have more area coverage, require half the volume of water, penetration is excellent and under surfaces can be sprayed. Many insects like thrips, jassids, aphids are available on the ventral surface of the leaves. In these cases power sprayer is very effective. In tea plant, stomata are situated on the lower surface of the leaves only. Hence, absorption is better with the help of power sprayers.
- An insecticide or foliar sprayer should not be used for spray of herbicides because the pressure required for former sprayer is 40 psi while for herbicides pressure requirement is 10 to 15 psi. If herbicide is sprayed through insecticide sprayers, discharge will be double.
- Spray calibration is extremely necessary to prevent wastage of costly pesticides and nutrients. It is necessary to put serial numbers in all the sprayers and check discharge rate of fluid from the nozzles every week. If the rate is more than prescribed volume replace swirl plates and discs. It is also necessary to find out the volume of fluid required to cover one acre area with individual sprayers. Accordingly dilution of chemicals and nutrients should be ascertained.

- In case of a cocktail, dissolve all the ingredients separately in a bucket one by one and transfer to the 200 litre capacity drum. Do not mix all the ingredients at a time.

Pest Control

- Pest surveillance is very important to monitor pests and design strategies to combat pest infestations.
- Colour chart on pest surveillance in each section should be maintained weekly. High attack can be marked with red and subsequent low degrees should be marked with yellow , blue and if there is no pest of a particular kind it should be marked green. This chart should be maintained for individual pests which are generally in that particular garden.
- Integration of synthetic chemicals, bio pesticides, cultural practices , genetic practices is extremely necessary to produce residue free healthy teas.
- Avoid using too much of synthetic chemicals , especially synthetic pyrethroids.
- Use right kind of chemical to control a pest or a combination of pest infestation. Wrong chemical can induce huge crop loss.
- Prophylactic spray of synthetic chemicals should be lessened as far as practicable.
- Pesticide spray should not be as a routine affair, it should be need based only.
- To control red spider mites, good shade tree stand, weed free sections, dust preventive hedges are extremely essential. Spray of Sulphur and Neem based products during January – February can be very useful in preventing large scale red spider attack.
- To control tea mosquito bug, it is necessary to adopt barrier method of spray and time of spray should be early morning and late evenings.
- Bio pesticides should be tried in between two spells of synthetic chemical spray. Bio pesticides are more effective when sufficient moisture is present in the atmosphere.
- It is better to use good new generation chemicals like flubendiamide, emamectin benzoate, abamectin, fenpyroximate, imidacloprid, propargite, thiamethoxam etc for effective control of pests. These chemicals can be little costly initially but are very much cost effective. You gain at the end.
- In the waterlogged areas, ground spray of *Trichoderma* has been found to be very effective.
- Spray of *Verticillium lecanni* , *Paecilomyces fumosoroseus* against red spider mites; *Beauveria bassiana* against thrips , jassids and tea mosquito bug; *Bacillus thuringiensis* against caterpillars and *Metarhizzium anisoplae* against termite and cockchafer can be very effective provided those are sprayed two to three rounds and proper spray technique is followed.

- Weed free sections and good drainage can reduce your pest problem manifold.
- Against red rust and die back, a combination of carbendazim and mancozeb can be useful.

NOTE: Above suggestions are completely based on the practical experience of the author and may or may not conform to the recommendations or specifications of any research center. Suggestions are mainly for the tea gardens of the plains of Dooars and Terai.

19

PHARMACOLOGY OF TEA

A CUP OF TEA

Anti-hypertensive	Anti-oral caries
Anti-diabetic	Anti-viral
Anti-obesic	Anti-halitosis
Radioprotective	Anti-tumor
Anti-ulcer	Anti-allergic
Anti-oxidative	Anti-amoebic
Molluskcidal Action	Fat Reducing Action
Deodorant Action	Anti-bacterial

Benefits of Drinking Tea

After Adam's wine water, Tea is the most widely consumed drink in the world. 3.5 billion cups of Tea is consumed daily, across the world . The medicinal use of tea was known long before it was used as a beverage. Tea is one of the most important non-alcoholic drinks and is popular due to its stimulating property. Tea was recognized for its medicinal properties in 2373 BC. Emperor Shen Nung of China was leading a caravan. On the way he became ill and arranged to take rest below a tree, which was incidentally a Tea tree. He wanted to have a hot cup of water. While the water was being boiled, a leaf from that Tea tree fell in the sauce pan . Shen Nung, after drinking that decoction, felt relieved and could experience the medicinal properties of Tea. The medicinal use of Tea is recorded in medical books of the Tang Dynasty (618 – 907 AD) in China , where it was noted that it has disease curing and detoxicating activity. With the advancement of science and technology , now it is confirmed that Tea is a store house of many medicinal properties.

Extensive research over the last few centuries throughout the world has confirmed that, apart from being a stimulant and refreshing beverage, Tea is also a health drink. Caffeine as a stimulant and Tea tannins or Tea polyphenols are the main constituents in *made* Tea. The polyphenol compounds are present as "catechins" in fresh Tea leaves or in green Tea and a part of them turns reddish "theaflavins" or "thearubigins" in the process of black Tea manufacture. .

Small amount of caffeine ingested per cup (40 mg) could only be beneficial as a stimulant. Tea possesses the property of stimulating formation of interleukin 2, which has important immuno-regulatory roles.

Black Tea contains a number of vitamins. Carotene, a precursor to vitamin A, has anti oxidant and protective properties. Thiamin (Vitamin B1) and ribo flavin (Vitamin B2) are both essential for releasing energy from food. Nicotinic and Pantothenic acids are necessary for release of energy from fat and carbohydrate. Ascorbic acid (Vitamin C) is essential for a healthy immune system. Vitamin B6 is involved in the metabolism of proteins. Folic acid plays a role in cell division. The average daily consumption of 3-4 cups of Tea provides 9 per cent of the daily requirement of Vitamin B1, 25 per cent Vitamin B2, 6 per cent Vitamin B and 10 per cent of Folic acid

Tea is rich source of Manganese and Potassium. Manganese is essential for the development of bones. 5-6 cups of Tea provides 45 per cent of our daily requirement of Manganese. Potassium regulates heart beats, it enables nerves and muscles to function and regulates body fluid. 5-6 cups of Tea provides 75 per cent of the daily requirement of Potassium.

During the International Tea Symposium on Tea and Human Health held in Sep,1998 at Washington, Director emeritus of American Health Foundation had quoted that regular intake of 3 or 4 cups of Tea can help in preventing heart ailment and various types of cancer.

Tea is a chemopreventive agent in cancer. Anti-oxidant properties in two cups of Tea are supposed to be equal to 3 to 4 glasses of orange juice or 20 glasses of apple juice. Tea extract containing chloroform, caffeine etc. has been reported to decrease mutagenicity of tumor. Certain phenolic compounds in liquor were claimed to be anti-carcinogenic because it is thought, they scavenge free radicals in cell cultures & the body and because they are good anti – oxidants. In Srilanka a single dose of black Tea extract containing 21.5 mg EGCG (epigallo catechin – 3 – gallate) and 22.7 mg of TF (theaflavine)] elevated serum antioxidant activity. Recent research findings indicate that tea may protect against the development of pancreatic and prostrate cancer. Scientists from the National Centre for Toxicological Research in the United States extracted theaflavins and polyphenols and demonstrated that they significantly inhibited the growth of human pancreatic and prostrate tumour cells. In another study, it was observed that the persons drinking two to three cups of tea a day can reduce the probability of prostrate cancer by 30 per cent .

Tea phenolics counteract with hypertension by depressing blood glucose levels .Cardiovascular diseases could be prevented through lowering of free cholesterol, total lipid and triglyceride levels in plasma and liver. By means of converting metabolic degradation products, Tea helps to maintain normal cholesterol level. Apart from this. Even Tea pigments noticeably reduce fibrinogen of patients suffering from cardiovascular diseases. Moreover, Tea polyphenols also strengthen the walls of blood vessels and regulate their permeability.

Tea extracts have also shown to be bacteriocidal e.g. against dysentery and in addition, extracts can also inhibit some bacterial toxins such as cholera for instance. Plant tannin in general, has been considered to have adverse effects on digestion or to cause constipation when consumed. In Tea, the scenario is different as because they are proved to have a variety of physiological beneficial actions that work to prevent the deterioration of foods, and prevent the infection of pathogenic bacteria and many diseases. The ratio of beneficial micro flora to harmful microflora determines the health of the digestive tract in humans. Tea drinking may play an important role in improving beneficial intestinal micro flora and providing immunity against intestinal disorders. Epigallocatechin gallate has shown very strong anti ulcerative activity. Hence, the misconception in the society that tea is harmful for stomach, it increases acidity, tea should not be taken in empty stomach and tea should not be provided to the children is totally baseless.

Both Catechins and the theaflavins derived from Tea can inhibit HIV – reverse transcriptase , thereby , reducing the incidence of AIDS.

Black Tea extract inhibits the oral fungus *Candida albicans* by 30 per cent. Its high fluoride content of 1 to 2 ppm can help in prevention of dental caries.

Dyspepsia and halitosis induced by smoking cause inconvenience and botheration. The food remaining in oral cavity during the ingesting process are decomposed into amino acids and are further decomposed into methyl mercaptan compounds. This is the main source of halitosis.It is also formed by the action between the nicotine and the proteins in the oral cavity Tea acts as deodorant . Chewing gums added with tea polyphenols have been developed in Japan and Korea for eliminating the unpleasant smoking halitosis.

Smoking is harmful to human health because tar oil is formed under high temperature condition during smoking. Research has proved that the harmful compounds produced in smoking process could be absorbed and degraded when the Epigallocatechin gallate or Theaflavin monogallate of Tea are added to the filter layer of cigarette.

Increased consumption of green tea has shown to be associated with reduced cell damage of the liver Catechin are used as liver protecting agent and it has shown a satisfactory therapeutic effect against chronic hepatitis in Japan.

Tea and Obesity

A human trial studying the relationship between green tea and body weight has found that men receiving catechin-enriched green tea showed a significant reduction in body weight, waist circumference and body fat as compared to men given catechin-depleted green tea. Average weight loss in catechin-enriched group was 5.3 lbs (2.4 kg), as compared to 2.9 lbs (1.3 kg) in subjects ingesting catechin-depleted product. In a separate study, scientists discovered that consumption of green tea catechins increased energy expenditure, by boosting the total amount of fat burned at rest and increasing the total number of calories burned on a daily basis.

White tea, made from the youngest and most tender buds of the plant, is capable of breaking down fat cells and preventing new ones from forming, according to latest research. Scientists from Germany tested extracts of white tea on human fat cells and found that they reduced the amount of genes associated with the growth of new fat cells and prompted existing cells to break down the fat they contain. In the US, a third of the population is categorized as obese, compared to a quarter in the UK.

According to a new Chinese study, funded by Lipton Institute of Tea, regular consumption of green tea was shown to promote weight loss and improve body composition in moderately overweight individuals. This study analyzed the effects of green tea consumption on body weight, body fat mass, as well as the distribution of fat. It found that body weight, waist circumference, intra-abdominal fat and the total lean mass all decreased after 90 days in the group that drank the tea with the highest concentration of catechins, equivalent to 6 to 10 daily cups of green tea.

Tea and Cancer

A new study by Japanese scientists has shown that green tea may prove highly beneficial for health and cut down the risk of developing blood cancer. Drinking 5 cups of tea per day was found to slow down the growth of cancer-causing viruses and micro-organisms. The study, carried out on more than forty thousand adults in Japan, suggests that those who regularly consume 5 cups of green tea have 42 per cent lower chances of developing all types of blood cancer than those who drink one or fewer cups of the green tea.

Scientists at the National Cancer Center in Tokyo examined six studies involving links between tea consumption and stomach cancer, and noted a 20 per cent decrease in stomach cancer risk for women drinking 5 or more cups of green tea per day. The results remain true, even when smoking was taken into account. However, the researchers found no correlation between tea drinking and stomach cancer among men.

Findings of a recent study into the effects of green tea on prostate cancer found that certain compounds have an influence on the progression of the disease. The studied substance, polyphenon, was found to significantly reduce the serum markers predictive of prostate cancer progression. In collaboration with Columbia University in New York City, the researchers are currently conducting a comparable trial among patients with breast cancer.

Scientists have discovered that a spray made from green tea could offer powerful protection against skin cancer. Researchers from Cleveland, Ohio say the tea spray has the capability to enhance skin's in-built immune system and reduce the damage caused by sun's ultraviolet light. It is believed that this blend can not only protect the skin, but prevent cellular damage that leads to skin cancer. Antioxidants known as polyphenols are thought to explain these beneficial effects.

Scientists found that high doses of a chemical found in green tea could kill cancer cells and reduce symptoms of leukemia, a blood or bone marrow cancer. Preliminary results suggest that humans are tolerant to very high doses of an active ingredient in green tea, epigallocatechin gallate (EGCG), and that many trial patients saw a regression of the disease. Researchers believe that the treatment would not be potent enough on its own, but could make other anti-cancer treatments like chemotherapy more successful.

Researchers at University Medical School in Arizona are looking for test subjects to participate in a study on green tea's potential in preventing the human papilloma virus, a virus that can lead to cervical cancer in women. Previous experiments have shown that green tea extract helps clear up cervical lesions created by the virus. The goal of the study is to see if more women who receive treatment with Polyphenol E, a chemical present in green tea, clear the HPV than those receiving the placebo.

According to latest research, women who drink a few cups of tea or coffee each day benefit from a lower risk of endometrial cancer, which arises in the lining of the uterus. Women who drank more than two cups of tea per day had a 44 percent lower risk of the disease, caused mainly by older age, obesity and higher levels of estrogen. Scientists point to caffeine as a possible explanation, as it induces certain enzymes that help neutralize potentially cancer-causing substances in the body. Underscoring this theory was also the absence of any link between decaffeinated coffee and cancer risk in the study.

Green tea catechins might strengthen metabolic defense against toxins capable of causing cancer, according to researchers at Arizona Cancer Centre. The study found that green tea catechin concentrate, equivalent to eight to sixteen cups of tea, boosted the production of a group of enzymes necessary to the body's defence against cancer-causing chemicals by as much as 80 per cent. It was the first clinical study that showed proof that catechins can increase detoxification enzymes in humans.

According to a Swedish study, women who drink at least two cups of tea per day have a lower risk of developing ovarian cancer. The study, carried out on 61'057 women aged 40 to 76, found that those who drank two or more cups of tea per day benefited from a 46% lower risk of ovarian cancer compared to non-drinkers. Furthermore, each additional cup was associated with an 18% drop in risk.

Tea and Bones

Scientists from Hong Kong have found that chemicals in green tea may help strengthen bones by stimulating their formation and slowing the process of weakening them. In a laboratory experiment on rat cells, they have noted that the ingredient epigallocatechin (EGC) in green tea boosted the level of a key enzyme that promotes bone growth by nearly 79 per cent. Also, they found that EGC increased the density of bone minerals in the cells, thereby reducing the risk of pain and fracture.

Tea and Heart Problem

A study from Scotland and France indicates that green and black tea may lower the risk of heart disease by preventing fatty deposits from building up in arteries. The research, carried out on hamsters with a high-fat diet, found a "remarkable" reduction in fatty build-ups of up to 96 per cent. Scientists believe similar effects can be replicated in human trials, as the administered amounts were equal to a cup of tea per day. The protective effect apparently comes from plant-derived chemicals known as phenolic compounds and flavonoids found in tea and some juices.

A recent French study suggests that women who drink three or more cups of tea per day reduce their exposure to heart attacks and strokes. The study examined the thickness of plaques in blood vessels, which are build-ups of fat or cholesterol and are linked to an increased risk of heart complications, and found those plaques were less frequent with increasing tea consumption in women. Strangely, no benefits were recorded among women who drank less than three cups of tea a day or for men.

Tea and Skin Aging

Latest study from Kingston University in the UK suggests white tea reduces age-associated wrinkles. Results showed white tea prevented the activities of the enzymes which breakdown elastin and collagen which can lead to wrinkles that accompany aging. The findings also indicate that white tea reduces the risk of inflammation which is characteristic of rheumatoid arthritis and some cancers.

Tea and Stress

Scientist found that tea-making can reduce stress levels by up to a quarter. The findings reveal that even a single cup of tea can significantly reduce anxiety levels after suffering a stressful experience. Moreover, psychologists discovered that the act of putting the kettle on also helped by tapping into a collective conscious and symbolism. The study showed a 25 per cent increase in anxiety for those that did not receive tea immediately after the stress-inducing test, whereas those who were given tea actually demonstrated a four per cent reduction in stress.

Tea and Diabetes

According to a study published in Journal of Food Science, black tea has the potential of managing the development of diabetes. Chinese scientists, examining the polysaccharide levels of different tea varieties, found that polysaccharides in black tea had the most glucose-inhibiting properties. Polysaccharides are a type of carbohydrate that may benefit people with diabetes because they help retard absorption of glucose. Black tea polysaccharides are also shown to have the highest scavenging effect on free radicals, which are involved in the onset of diseases such as cancer and rheumatoid arthritis.

Scientists at the Medical College of Georgia found that green tea antioxidants may slow or even prevent the development of type 1 diabetes. The study, carried out on laboratory mice, tested the effects of green tea's predominate antioxidant, known as EGCG, on type 1 diabetes and Sjogren's syndrome, which damages moisture-producing glands causing dry mouth and eyes. It was discovered that the antioxidant compound reduced the severity and delayed the onset of salivary gland damage associated with Sjogren's syndrome - a condition with no known cure. EGCG also dramatically slowed the development of type 1 diabetes in the rodents.

Latest research from the University of Dundee, Scotland suggests that black tea may help prevent type 2 diabetes, which develops when the body fails to produce enough insulin. The researchers have discovered that several components of black tea, namely theaflavins and thearubigins, mimicked insulin action and could have beneficial effects on the most widespread form of the disease.

Tea and Strokes

In a new study, Australian scientists have found that frequent tea consumption can significantly reduce the risk of suffering a stroke. Researchers discovered that long-term consumption of green tea, at least once a week, substantially decreases ischemic stroke risk. The study looked into green tea consumption of Chinese patients who suffered a stroke and found encouraging news when comparing frequent drinkers with infrequent or nondrinkers. Major inverse dose-response relationships were also found for years of drinking and the amount of dried tea leaves brewed.

A new study by a group of Swedish scientists suggests that coffee and tea consumption may reduce the risk of stroke caused by a blockage. The study, carried out on over 26,000 Finnish male smokers, examines ties between diet and risks of different types of strokes caused by bleeding in the brain and artery blockage. The results show that men who drink two or more cups of tea per day have a 21 per cent lower risk of this type of stroke compared to non-drinkers. Scientists suggest tea's phenolic compounds may be instrumental in preventing cerebral infarction and hope to confirm the results among women and non-smokers.

Among many other steps and practices, the National Center for Chronic Disease Prevention and Health Promotion (a branch of the Centers for Disease Control), recommends people trying to quit tobacco drink tea instead of coffee.

Tea and AIDS

A green tea compound is found to inhibit sexual transmission of HIV virus when used as an ingredient in vaginal creams. German researchers discovered that the green tea polyphenol, called epigallocatechin-3-gallate (EGCG), is capable of neutralizing a protein in sperm which serves as a vector for viral transmission during sex. Scientists believe that EGCG could provide a simple and affordable prevention method for poor and developing nations, and would also enable women to ensure proper protection during sexual intercourse.

Tea and Dental Diseases

New research suggests that green tea can improve the health of teeth and gums. The study, carried out on 940 men in Japan, found that people who regularly drank green tea had very healthy teeth and gums, compared to those who did not routinely drink it. According to research, the anti-inflammatory effects of green tea catechins reduce the risk of periodontal disease - less bleeding with probing, decreased depth of periodontal pocket and decreased clinical loss of attachment of tooth to gum.

A study on the effects of drinks on teeth has found that brewed tea possesses much less negative erosive properties compared to other soft drinks. According to research published in General Dentistry journal, tea poses virtually no risk to the hard part of teeth, or enamel, whereas refined sugars and acids found in soda and citrus juice promote tooth erosion, which is irreversible. Academy of General Dentistry spokesperson has therefore highly recommended patients to choose tea as an alternative to more erosive drinks like soda and fruit juice.

Tea and Brain Performance

According to new research, people who consume tea, wine or chocolate have higher brain performance than those who do not. An international team of scientists examined the relation between cognitive performance and the intake of those three substances. They found that those who consumed chocolate, wine, or tea had significantly better mean test scores and lower prevalence of poor cognitive performance than those who did not. The active ingredients studied were flavonoids, micronutrients found in plant-derived foods.

Tea and Athletes

A Brazilian study measuring the effects of green tea on athletes undertaking resistance training indicates that its consumption increases body's antioxidant defense capacity. Researchers say that if the results can be confirmed on a larger scale, the antioxidant-rich beverage could be proven valuable in fighting oxidative damage caused by exercise. The findings may pave the way for green tea to gain access to the energy drinks market, which is valued at £137 million in the UK and almost $3 billion in the US.

Tea and Antibiotic Efficiency

Researchers in Egypt found that green tea enhances the effects of antibiotics up to threefold in fighting drug-resistant bacteria. The study, carried out by the Alexandria University, suggests that in almost every case and for all types of antibiotics they tested, drinking green tea at the same time as taking the antibiotics appeared to increase the action of the antibiotics and reduce drug resistance in bacteria.

Tea and Bladder Inflammation

Preliminary findings by the University of Pittsburgh School of Medicine suggest that components in green tea may protect bladder cells from inflammation. In the experiment, normal and cancerous bladder cells treated with green tea catechins showed signs of increased protection against hydrogen peroxide, which damages or kills cells. About ten million Americans suffer from various forms of bladder disease.

Tea and Rheumatoid Arthritis

A study by the University of Michigan suggests that green tea could ease rheumatoid arthritis by blocking molecules that break down bone in joints. Scientists noted a significant suppression of activity in molecules that cause joint inflammation after exposure to EGCG catechins. While not conclusive, the study indicates that green tea may become a natural alternative to chemical treatments for about 2.1 million Americans who suffer from the disease.

Tea and Bacteria

Researchers from Slovenia have found that green tea helps prevent bacteria. According to the study, green tea catechins inhibit essential bacterial enzyme DNA gyrase, thereby offering an alternative to compounds that have side-effects. This finding, the scientists say, may be used to develop even more potent antibacterial compounds based on EGCG.

Tea and Radiation

A group of American and German scientists have discovered that tea extracts offer an effective treatment of skin damaged by radiation in cancer therapy. The study found that tea extracts inhibit inflammatory pathways, thereby reducing inflammation, and reduce the duration of skin damage caused by radiation treatment by up to five to ten days. Researchers noted that green tea showed higher anti-inflammatory properties than black tea in human white blood cells.

Tea and Influenza

Biochemical studies of green tea are beginning to show anti-influenza effects. These findings indicate that this is mainly due to a specific polyphenol found in the tea leaf called EGCG. The studies show that EGCG has potentially direct virucidal effect on influenza and possibly odther viruses.

Thus, it can be said that few cups of tea a day can keep the doctor away.

Colour Plates

Camellia sinensis (China variety) (Page-4)

Camellia assamica (Assam Variety) (Page-5)

Camellia sinensis ssp lasiocalyx (Cambod variety) (Page 5)

Anthocyanin pigmentation in Cambod variety (Page-6)

Anthocyanin pigmentation at the petiolar end (Page 6)

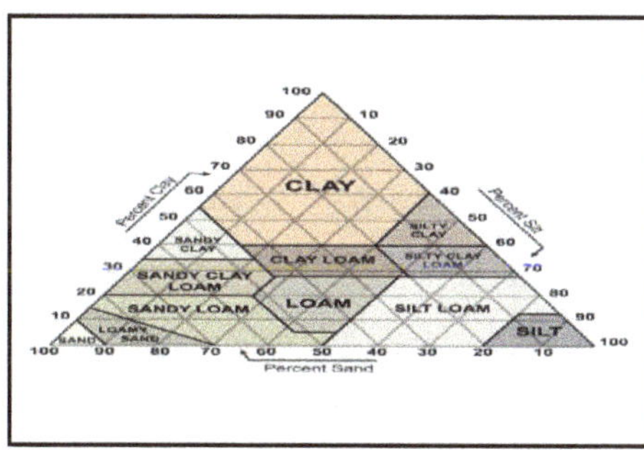

Soil textural classification (Page 29)

Zinc, Magnesium, Nitrogen and Potassium deficiency (Page 47)

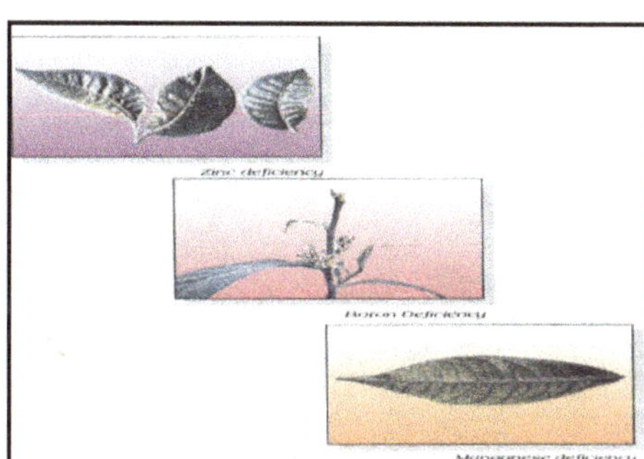

Zinc , Boron & Manganese deficiency symptoms (Page 48)

Acute Magnesium deficiency (Page 48)

Vegetative propagation (Page 62)

Netlon shade (Page 64)

Arched transparent polythene tent (Page 65)

Young tea plantation (Page 69)

A Sub-Main drain (Page 79)

Water hyacinth (Page 92)

Ageratum conyzoides (Weed extract for pest control) (Page 105)

Artemisia vulgaris (Weed extract for pest control) (Page-105)

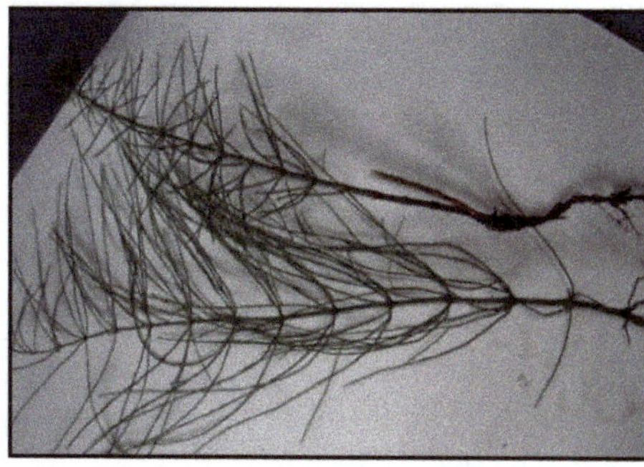

Equisetum arvense (Weed extract for pest control) (Page-105)

Urtica dioecia (Weed extract for pest control) (Page-105)

Polygonum runcinatum (Weed extract for pest control) (Page-105)

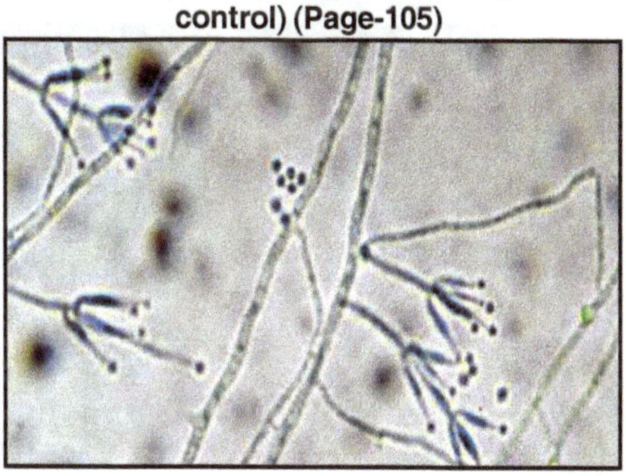

Paecilomyces fumosoroseus (Page-114)

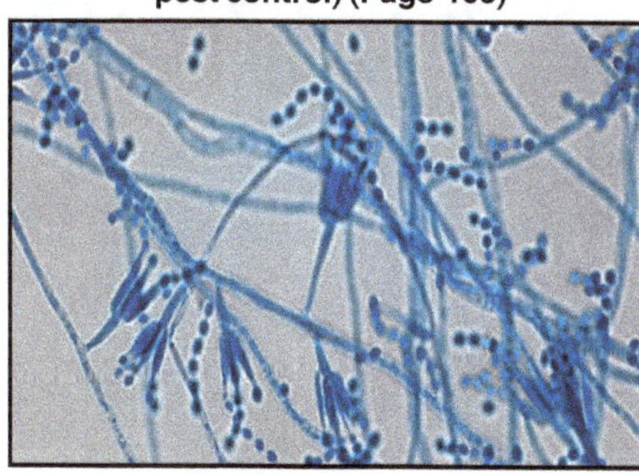

Paecilomyces lilacinus (Page-115)

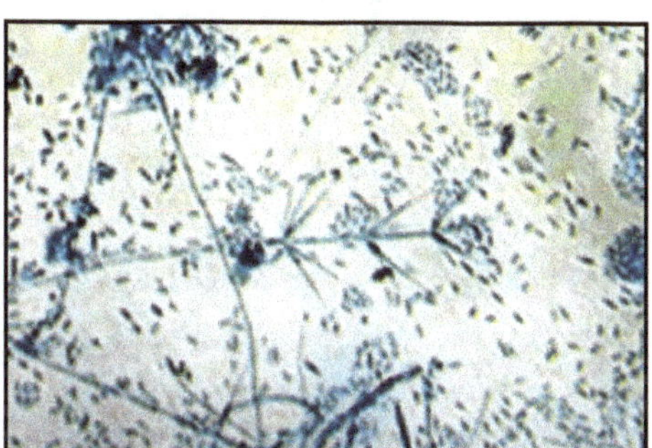

Verticillium lecanii (Page-115)

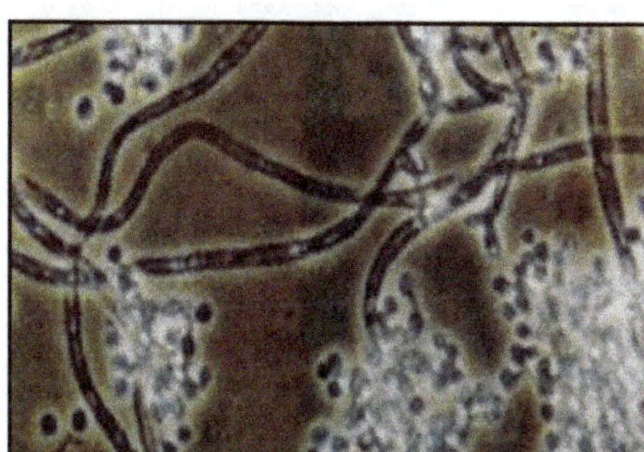

Beauveria bassiana (Page-115)

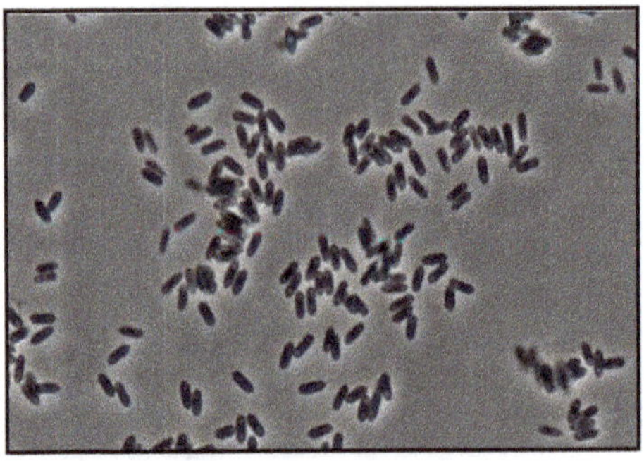

Metarhizium anisopliae (Page-116)

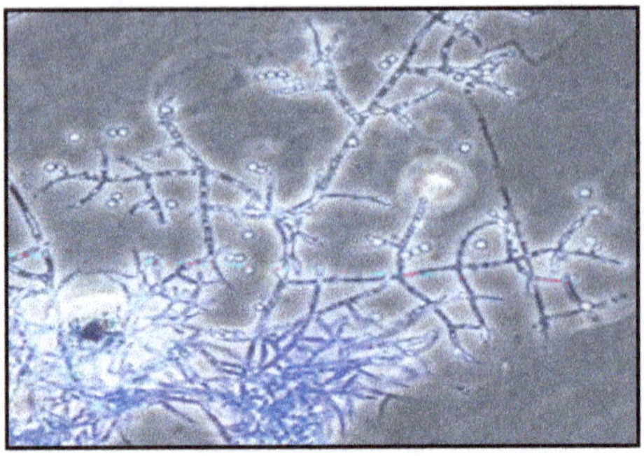

Trichoderma viride (Page-116)

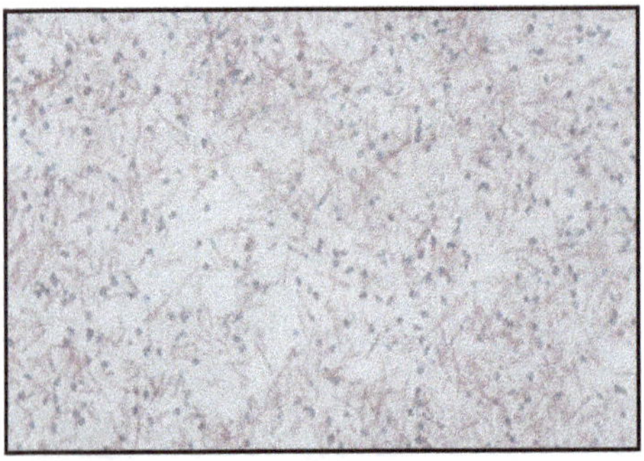

Pseudomonas fluorescence (Page-117)

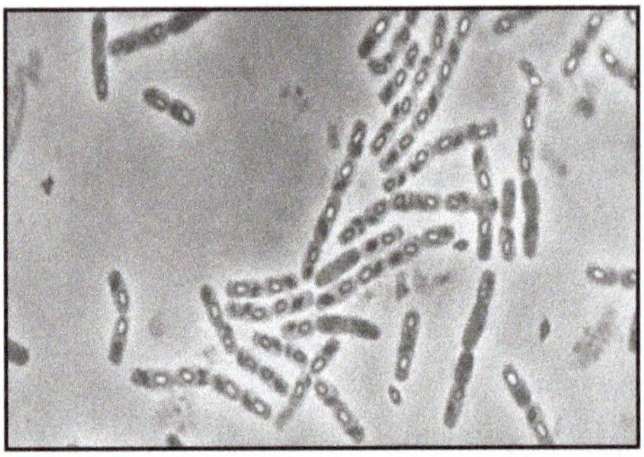

Bacillus thuringiensis (Page-117)

Eudrilus eugeniae (Page-124)

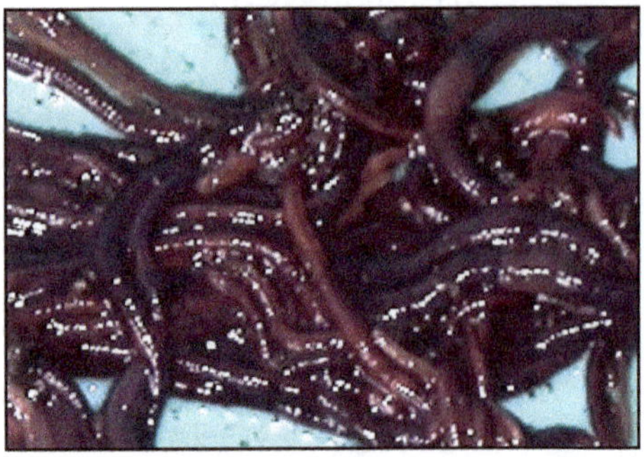

Perionyx excavatus (Page-124)

Eisenia foetida (Page-124)

Amyanthas diffringens (Page-124)

Albizzia chinensis (Page 127)

Albizzia odoratissima (Page 128)

Albizzia odoratissima saplings in the nursery (Page 128)

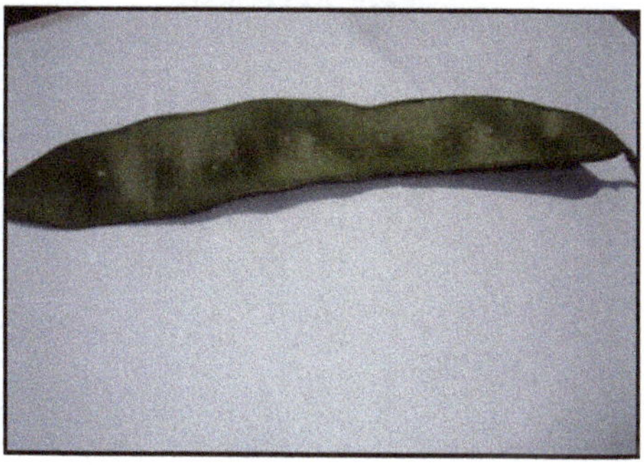

Albizzia odoratissima pod (Page 128)

Pods of *Albizzia lebbek* (Page 130)

Cockchafer grub (Page-157)

Termites (Page-158)

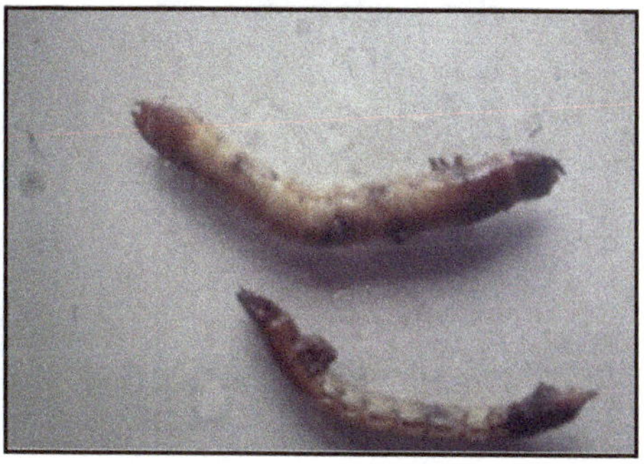

Red Borer (*Zeuzera coffeae*) Page-159

Light Pruning (LP) (Page-137)

A pruned tea section (Page-139)

Medium pruning- big cut marks with bitumen paint (Page 139)

Deep skiffed tea section (Page 140)

Medium skiffed tea section ((Page 140)

Levelling off skiff (Page 142)

Breathers (Page-143)

First FFP (Page 143)

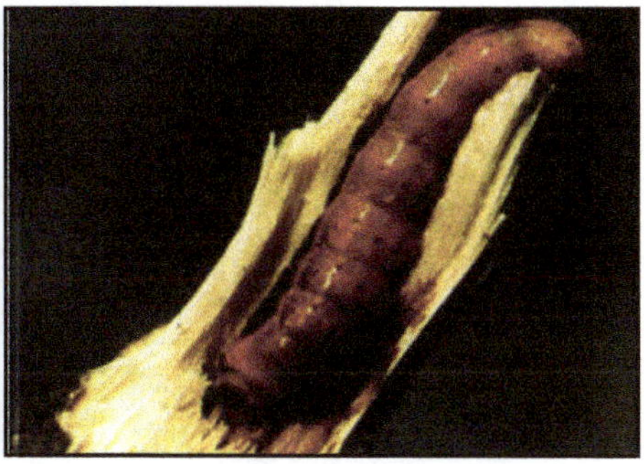

Red borer inside a stem (Page 159)

Moth of red borer (Page-159)

Tea mosquito bugs (Page-160)

Tea mosquito bug punches (Page-160)

Leaf crippled after severe attack of tea mosquito bug (Page-161)

Tea thrips (Page-160)

Sand papery lines on the leaf after Thrips attack (Page 161)

Tea Jassid (Grren fly) (Page-162)

Jassid infested tea plant (Page-163)

Looper caterpillar infested leaf (Page 167)

Moth of *Biston (=Buzura) suppressaria* (Page-167)

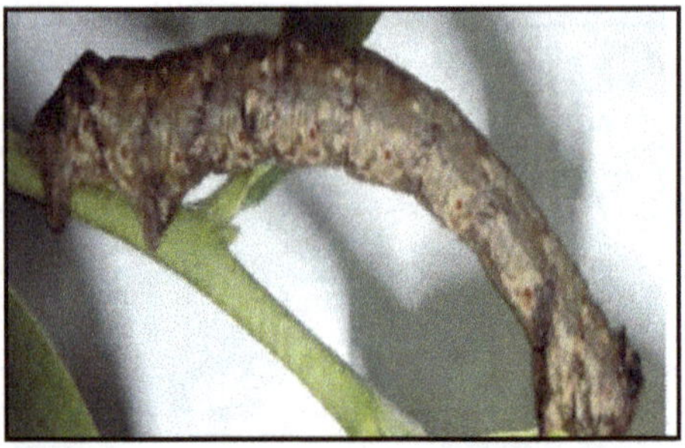

Larva of *Biston (=Buzura) suppresaria* (Page-167)

Moth of *Hyposidra talaca* (Page-168)

Moth of *Hyposidra infixaria* (Page-169)

Larva of *Hyposidra infixaria* (Page-169)

Moth of *Ecctropis* sp. (Page-169)

Larva of *Ectopis* sp. (Page-169)

Moth of *Ascotis* sp. (Page-170)

Larva of *Ascotis* sp. (Page-170)

Moth of *Cleora* sp. (Page-170)

Larva of *Cleora* sp. (Page-170)

Red Slug caterpillar (Page-171)

Leaf infested by Red slug caterpillar (Page-171)

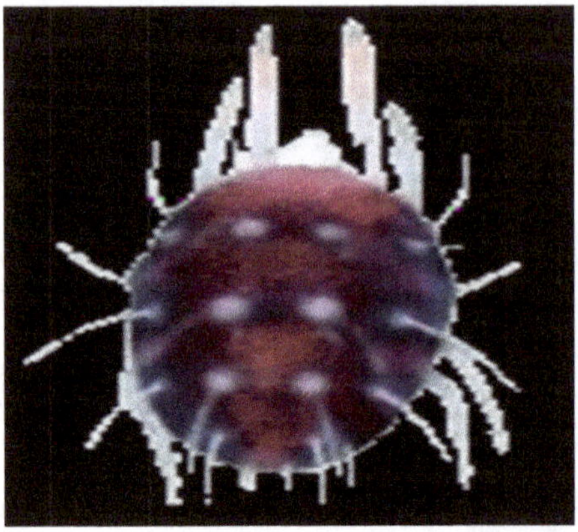

Red spider mite (187)

Red spider mites on the leaf surface (Page 187)

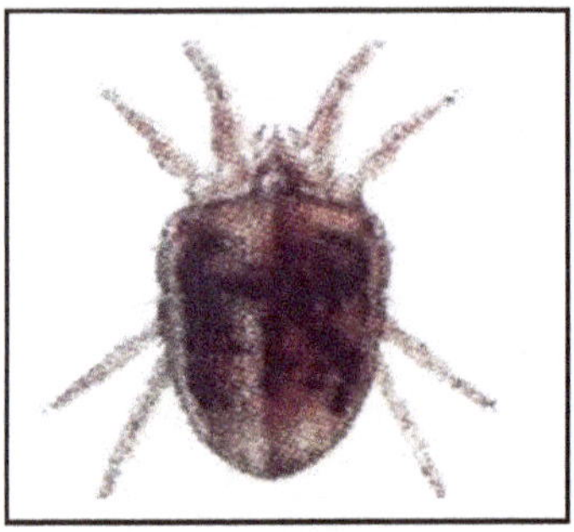

Scarlet Mite (Page-188)

Pink Mite (Page-188)

Purple Mite (Page-189)

Brown Root Rot (Page-199)

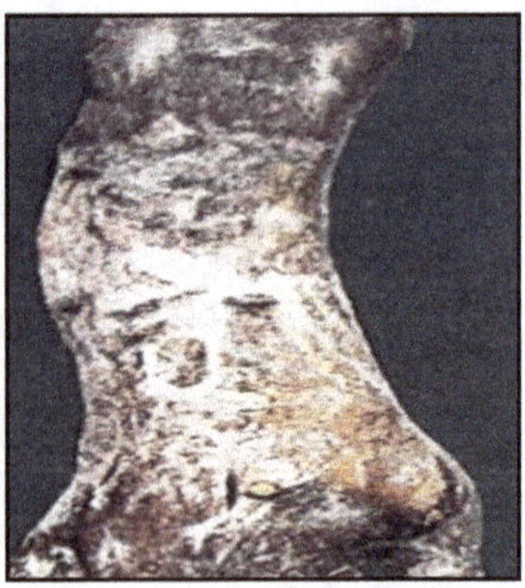

Red Root Rot (Page-199)

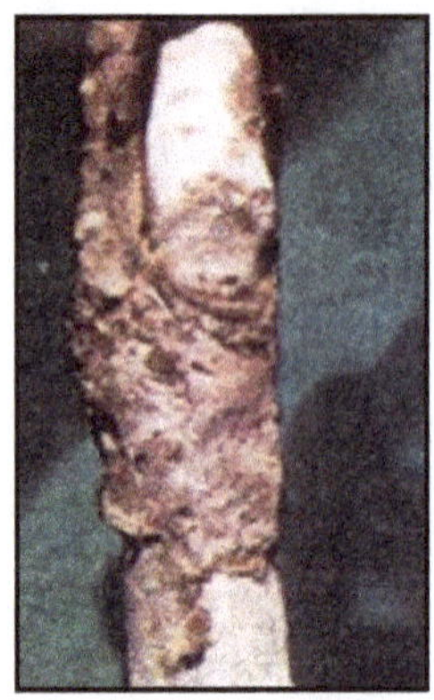

Purple Root Rot (Page-201)

Charcoal Stump Rot (Page-202)

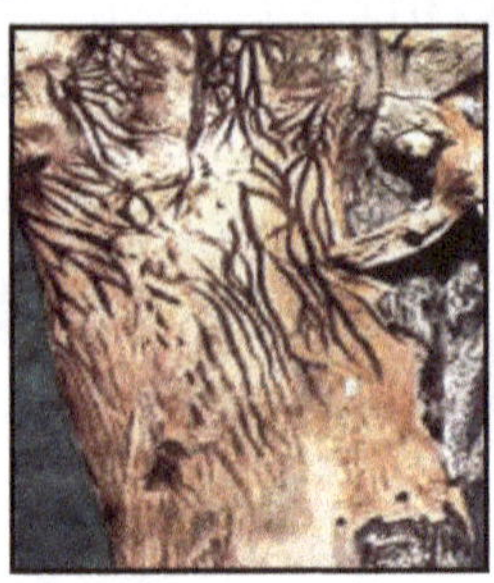

Viole Root rot (Page-203)

Stem Red Rust (Page-203)

Black Rot of Leaves (Page 204)

Leaf Red Rust (Page-205)

Brown Blight (Page 206)

Grey Blight (Page 206)

Blister Blight (Page-205)

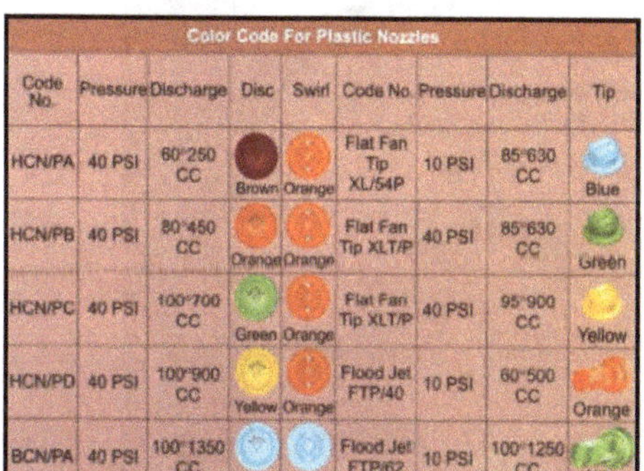

Color Code For Plastic Nozzles								
Code No.	Pressure	Discharge	Disc	Swirl	Code No.	Pressure	Discharge	Tip
HCN/PA	40 PSI	60-250 CC	Brown	Orange	Flat Fan Tip XL/54P	10 PSI	85-630 CC	Blue
HCN/PB	40 PSI	80-450 CC	Orange	Orange	Flat Fan Tip XLT/P	40 PSI	85-630 CC	Green
HCN/PC	40 PSI	100-700 CC	Green	Orange	Flat Fan Tip XLT/P	40 PSI	95-900 CC	Yellow
HCN/PD	40 PSI	100-900 CC	Yellow	Orange	Flood Jet FTP/40	10 PSI	60-500 CC	Orange
BCN/PA	40 PSI	100-1350 CC	Blue	Blue	Flood Jet FTP/62	10 PSI	100-1250 CC	Lime

Description of Plastic nozzles (Page 238)

Spray operation in progress (Page 241)

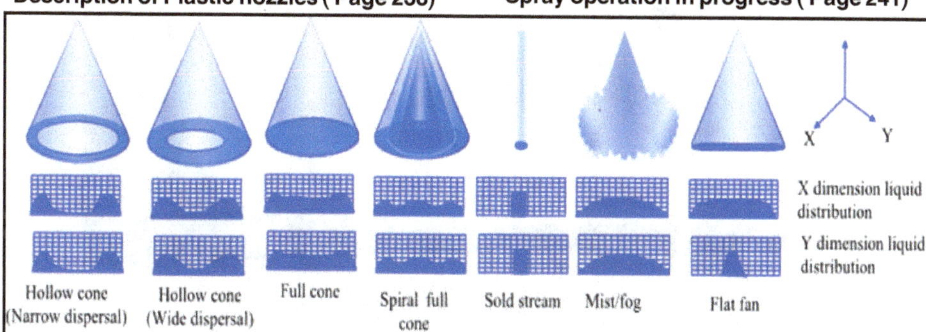

Diffrent types of Nozzels (Page-238)

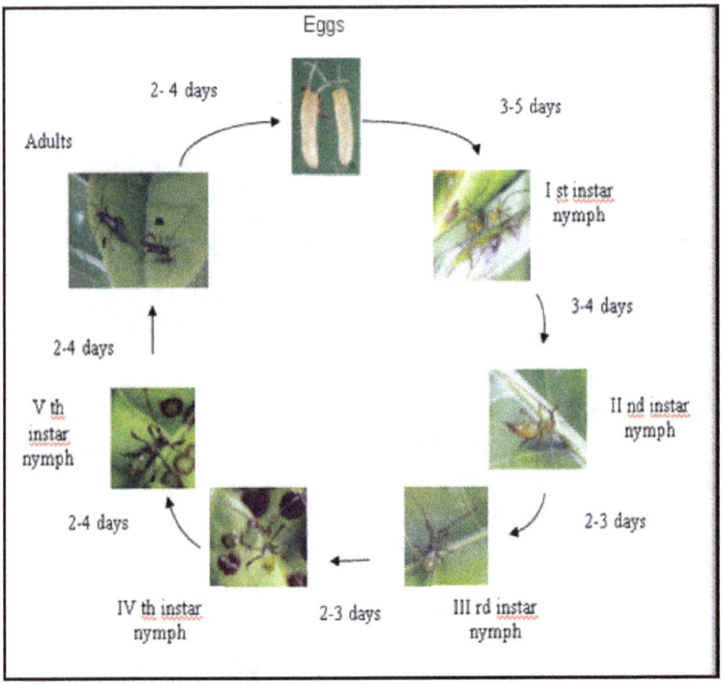

Life-cycle of Tea Mosquito Bug (Page-194)

Life-cyle of Thrips (Page-194)

www.ingramcontent.com/pod-product-compliance
Lightning Source LLC
Chambersburg PA
CBHW071233230926
51849CB00001B/90